GARY AND DUKE

A Totally True Story

NED McFADDEN

Copyright © 2021 Ned McFadden
All rights reserved
First Edition

PAGE PUBLISHING, INC.
Conneaut Lake, PA

First originally published by Page Publishing 2021

ISBN 978-1-6624-3092-3 (pbk)
ISBN 978-1-6624-3093-0 (digital)

Printed in the United States of America

Dedicated to Chester "Pop" Sandford
Acknowledgments to Celestine Fitzsimmons Gookin,
Marlene Dowd McFadden, and Sandra Lupo McFadden

CONTENTS

Prologue: Bug-Eyed 7
Introduction: Bugs Funny 11
Chapter 1: Two Healthy Teens from the 'Burbs......................15
Chapter 2: Too 19
Chapter 3: Introduction to Big Sur 23
Chapter 4: Three Thousand Dollars' Worth of Drugs..............28
Chapter 5: Thermometer Outside..33
Chapter 6: He's a Dead Head...39
Chapter 7: Pancake Makeup 43
Chapter 8: Sunset Strip 52
Chapter 9: Jack and Jay...56
Chapter 10: Headed East...60
Chapter 11: Rocking in the Sixties..68
Chapter 12: Hitchhiking Some 77
Chapter 13: Crossing Missouri 82
Chapter 14: No Soap 93
Chapter 15: Echo Chamber 96
Chapter 16: Nearly Finished 104
Chapter 17: Out-Duked 112

PROLOGUE

Bug-Eyed

The sedan was idling on a dirt roadside stretch, just off to the side of a two-lane road inside South Caroline. Gary told me, "It's your turn to drive." In my first attempt at a stick shift, I went to turn off the radio. The car had no radio. I tried to engage gears, but the engine stalled. When I tried again, the car lurched forward over a curb, and the tailpipe fell off Gary's emerald-green MG sedan. His lips pursed while he said nothing. Gary exited the car and walked away. For fifty minutes, I contemplated what Gary would say when he returned. Gary had found me some telephone wire, which he jury-rigged about the car to last several more miles.

We were seventeen and told our parents we were snow skiing in New Hampshire, north of our home in Wellesley, Massachusetts. We were about to land in Miami Beach as high school seniors for Christmas recess. Sleeping, we scrunched upright in the car. We ate breakfasts opposite the South Beach—the two eggs special with coffee, toast, and a small orange juice for twenty-nine cents.

"Can't sleep on the beach," the cop said, kicking my naked heel.

"Can't sleep in a cell," another more affable cop said later at a precinct station. "It's New Year's Eve." We watched the Orange Bowl Parade that included my missing a date and having my first dry shave. For an entire week, we wrestled ourselves silly in the Atlantic and didn't eat well.

"It'll be nice getting back to school for some rest." We laughed.

One particular morning, I was leaving the surf, but Gary was not right behind me. He was out in the deeper water, hands on hips. I ran for the beach, went in a way, lay down, and studied Gary Sandford being pensive. He walked out of the water, looking downward, not noticing how his two feet leave a single wake. With a fixed stare, Gary crossed the beach and headed for a telephone booth next to the road. I soundly punched the sand, knowing my friend is going to call home. I went over by the road, my mind racing what to do or say. Looking in every direction and finally up, I saw atop the well-worn weather-beaten phone booth a dragonfly that seemed to be eyeballing me. When I noticed it a third time, I thought, *Could be.*

"Hello, Sue. Put Dad on the phone," Gary said. I was paying attention to that darn dragonfly on the top of the phone booth. I contemplated entomology.

"Hello, Pop, we're in Miami Beach," Gary rejoined. "What? Yeah, yeah, the Miami Beach." Gary "sheeeshed" and rolled his eyes up.

I swear by this…the dragonfly took off and came for me square between the eyes, hovered in front of my schnozz, buzzed a circle first completely around my right ear and then my left, went back in front of my face, and then returned to its perch. I started thinking, *Dear God, is it possible that insect is with some kind of message?*

Then an inner voice beaconed to me, "You think Gary is tired? Wait until you finish chronicling what happens to you two next."

"Boy," Gary said crisply, "you should have heard how angry my father was, I mean is."

I ceased holding onto the message that was given to me. I succumbed to Gary's presence, saying, "Thanks a lot. Now your folks will call my parents."

"Anytime, old boy," Gary responded. He didn't need any drop to get the better of me in any punch out.

Prior to the drive home, oil and gas money were earned by Gary, climbing up the coconut trees and tossing green coconuts down to me. These we sold to tourists for fifty cents apiece while a ripe one in a supermarket cost a quarter. Because of the tailpipe incident, Gary drove the entire length of the road back to Massachusetts despite the worst snowstorm in two decades. Inside the Holland Tunnel, he fell asleep at the wheel and was heading into a yellow checker taxi, but I straightened the wheel, and he resumed driving, not knowing that he had been driving the MG while asleep. I must confess that after the trip, I turned up at school, having used a sunlamp, and my face was a burnt red.

"Where we going next?" Gary asked me when we finally bumped into one another in school.

"What do you mean?" I asked.

"After graduation, I hear Bob Saveland and Phil Santuchi are already planning a trip to Europe. We gotta do something to top them. You don't know them like I do. I live next to them. We've gotta beat them."

INTRODUCTION

Bugs Funny

Early on a summer, moonlight shone through the venetian blinds, and I woke up with a smile on my face. I stared at an aged-beige metal flaked radiator whose steam valve has been missing for years. Instead of sighing, I drew in a gasp of air as I recollected the way steam used to shoot out of the radiator, and I'd wake up to a room full of steam. I realized that within twenty-four hours, I was going to be at least five hundred miles away, and I didn't know in what direction. Sucking in the comfort of the covers, I rose to dress in the day's bright air. Flying down two flights to the English Tudor home, I paused at the kitchen picture window. There separating houses from our neighbors, the Flynns were neatly rowed pine trees being warmed and nourished by the sun. I paused for an inner private moment. Then reaching for the telephone, Gary was on the other end without a ring having sounded.

"Hi, ya ready to go?"

"Told you I would be," I spoke.

"Where are we going?"

"Look, Gary, tell you what. I've gotta eat breakfast. I'll call back in fifteen minutes. But just think about New Orleans, okay?"

"Make it ten," Gary said, blanking the call.

In the kitchen was Mother making breakfast. She conspicuously minded her own business. I started in on breakfast. She stood in such a special way that I could tell she's thinking very hard. I looked at the sun's reflection off the kitchen round wooden table finish at that reflected white orb I stare. I became halted in my thoughts when mother plainly said, "Why don't you go to California?"

I found myself masticating, looking closely at the finish on the table at my knife, fork, and spoon. *California*! My mind shouted as I stared at my eggs disinterestedly. Images of beaches, waves, girls, gold, deserts, especially blond girls, leaped into my mind. I stood and went over to my mother and kissed her cheek. She wore a mixed expression.

"Hello, Gary, me here. I say Cape Cod is too close to home."

"Really, Sherlock. Sure glad you handled that problem."

"Cut it out wisely, guy," I snapped back.

"So what do you want to do?"

"Well, I was thinking New Orleans, but my mother suggests California."

"Okay, then we'll take the continental trip. When can you come over here?"

"I'll be there in two hours, at ten."

Next thing going on was my mother. Marie McFadden was on the phone with Carol, Gary Sandford's mother. Their talk was intriguing while I was reflecting on how I might meet my future.

"Mrs. Sandford and I will each pay for a one-way ticket for each of you to San Francisco."

California—it was better than we imagined.

I made excuses about errands and borrowed the car to go over to Martha Chandler's home.

"She'll still sleeping, but I think it's all right if you go up there. I understand you'll be traveling far. Good luck," Ms. Chandler spoke to me person-to-person, which I appreciated. Martha's face was covered by her sleek black hair. When the curtains were drawn, she kissed back, not even opening her eyes.

"Mmmm, I just knew you wouldn't leave without seeing me one more time."

Gary Sandford toted the nickname of "Brute Force." This came from an incident where he put an older collegians head to the ground while pumping from the ankles. Gary was a truck driver's sort of man—robust, hearty, almost burly while not just eighteen, an absolute enjoyer of life who could have a hearty laugh whenever he feels the situation requires such. He was five-foot, ten inches one eighty-five pounds. He possessed a soft while commanding voice. He had green eyes that were often observed. He fit into a crowd and liked it that way. Often, he'd practice lessons at night away from school—the punching bag and weight lifting.

When Gary appeared at 524 Worcester Street at ten o'clock, he carried what sort of a cross between a sewing machine size traveling bag and a leather carpet bag, which leather winkles were well worn and was toted by a temporal-looking handle is.

"What about the handle, man?" I asked.

"Hey! Don't bug me. You just worry about yourself."

An often worn phase in the Sandford household was "It's just Gary."

Mom took Gary from the kitchen into the dining room, ignoring to close the two-way swinging door. My eyes widened as I saw her hand a small wad of bills to Gary, saying, "Promise me you'll take care of Ned."

"You're going to have a worse time than I am." Gary zinged me as he pinched his stomach bulge. I was seventeen in 1968. Our hair color was dirty blond, although mine was cropped more closely to prevent the reign of curls, which was not stylish. I weighed one hundred sixty pounds from several seasons of cross-country and two-mile track running. My eyes were sky blue, while Gary's eyes were hazel. The occasional person thought we were twins.

In my fourteen-square-inch army green canvas bag, I carried no razor, but I did carry a small kit for abrasions, flashlights, three pairs of U-trou, three colored T-shirts, three pairs of socks, and a gold striped purple football shirt. Inside the safety kit were three phone numbers of never met relatives in Pittsburgh, San Francisco, and Los Angeles. I wore leather boots, a pair of dungarees, and a leather jacket.

"How much money you got?" Gary asked me eye to eye.

"About ninety, you?" I replied.

"None of your business."

We missed the first flight out, and Mom tweaked my side, which made me want to cry. Gary said, "Boy is she smart seeing us off to make sure we buy the tickets."

Halfway through the flight, a movie was shown onboard. It depicted an interracial couple who were about to wed, meeting their respective parents with there at first and then later reaction being the point to the story. When the end came and the parents were to try, the elderly Black lady to Gary's left was crying and blowing her nose. Gary was stuffing his fist into his mouth, not crying. Soon the stewardess proffered a fruit bowl of plenty to the passengers, and she returned to Gary and his schtick. Another full bowl was offered and just about all taken and put into this unbuttoned shirt.

We arrived at San Francisco's airport at night. I eyed closely, oh so closely—three flights of metal stairs that led to a limousine stand. Gary kept tugging at my elbow, which made me think he wanted our first fisticuffs. Only he left me and later returned with a dark-skinned chauffeur who is going to drive us into the city (Oakland) for a total price of ten dollars. As we drove into Oakland proper, I noticed a number of massive industrial buildings, which struck me as America's wealth suppressing the populace. We drove onto a street so dark that I checked to see if I have my sunglasses on. The house that we stopped at was one story, had a condemned notice affixed to the front of the property, and had wooden boards over the windows.

"Stay here," the driver said. Then he continued, "There's nobody there."

Gary marched up a cracked sidewalk along with an overgrown brush and got to the unlit house where he knocked loudly on the door. A dim light went on, and mere seconds, the front door creaked open, and Sue Sandford leaped into her brother's outstretched arms. Sue spotted me and whispered, asking Gary for my name. Then she said, "Ned, of course."

The inside was a dark ramshackle invitation for ghosts and insects. There were two bedsheets covered in two dimly light shades. Sue had thought evictors were at the door. We met a fellow who would turn out as Sue's husband, Tom. The night commenced with we two boys sleeping in opposite armchairs, and once the lights were off, the insects came to play. The bugs won.

CHAPTER 1

Two Healthy Teens from the 'Burbs

After sharing bowls of cereal with Tom and us two guys, Sue told us to skedaddle, and I was relieved that we were going.

"Be cool," Sue told us. "Act like you're spaced out on drugs or something. Be weird. Most of the folks here are Black and out of work. And they won't care about you one way or another. But some are looking for trouble even first thing in the day. Now don't give me that look, Gary," Sue spoke to her brother's taller chin. But really, Gary and I had no concept of what our behavior should be out on the street.

You all know that to get to the province of San Francisco from Oakland, one necessarily crossed the Oakland Bay Bridge—a magnificent structural steel creation. There was a beautiful scene of boats, motorboats, sailboats, ferries, and ocean crests. The water capped where the wind gusted, and the glistening white flume glistened in the morning sun. The warm sea air made me light-headed, and the view combined with San Francisco's glass building reflecting the sun made the view magnificent. The sun, which in just days would be coming pell-mell, it lit so bright and cleanly that by it alone was known. "T' will be a full day."

Striding the sidewalk, we were drawn into a downtown cavort as if with a course well defined. Ladies in colorful prints smiled, as businessmen nodded "hello." After realizing that we have no idea where we were walking, we stopped at a restaurant for minutes and watched through a plate glass window at a man eating his breakfast. Then we decided to separate, and each one of us would find work. After agreeing to meet at noon at a nearby coffee shop, we each entered the office buildings opposite one another.

"No permanent address in California, no," said a comment I met by on three different floors of a thirty-story office building. I didn't even have a copy of my high school diploma. Into the coffee shop, I sauntered. Black. Moments later, Gary walked into the restaurant and took the stool next to me.

"Maybe," I said. "I was raised on pure energy, but let's find a place to go swimming in the ocean."

"Gotta find a cave first," Gary said. "Besides, we're in the inner city."

Every step, we marched tenderfoot, naive, careless, or just stupid. We didn't think of rooming a house. After the whole day, we were tired and no better off. We walked the most crooked street in the world. Up it! The streets were very, very steep…once I had to check myself from falling topsy-turvy over. From the California street top, where the cable car operates, we saw off near the coast many, many paces away from a tower. To this, I wanted to hike.

"Will be dark by the time we get there," my partner disagreed.

Telegraph hill, the tower atop the steep hill, was the answer to our prayers. The shadows of the night didn't deter us from passing through thick brush as we climbed the hill instead of walking the most lengthy driveway. Climbing the stairs to the tower, one inevitably said, "We're getting as high as possible in San Francisco without any drugs." From the observation's deck, one saw the seven hills of San Francisco shinning up interestingly. Then seaside, Alcatraz, three-quarters of a mile out to sea, was ominously alone. Also, inside the observation deck were three steps leading to a wood and wire mesh door. We rammed at the door, but it pulled to open.

Twelve months earlier, we stood three floors up inside of Wellesley College's tower facing similar three steps and a wood and wire mesh door. Unable to get inside, we walked away, frustrated. On the ground level, we saw carpenters working, leaving their tools scattered. We swiped a hammer, chisel, and wire cutters. Soon we bounced shoulder to shoulder off the doorframe leading onto a cavernous steeple room at least seventy feet in height. There was a platform leading to a musical organ-looking device. There as well were four corner towers, each protected by lots of dust and cobwebs. Gary climbed up the flights of stairs to get to where the organ-like console was…for pealing the tower's bells. I raced upward atop one spire, pushing away cobwebs and dirt. The tower bells beginning to peal most loudly prompted me to race for the summit. I raced toward the top. Then free at the summit, hanging down from the top with tape, was a woman's sanitary napkin.

Running down the hundred or so stairs to Telegraph Hill's tower, we raced without concern for any people possibly walking up. The sound of our feet hitting the pavement clashed in the night. Taking the hill down, when I twisted my ankle, I thought how I'd never again cross this hill.

The morning brought hundreds of cars and vans and trucks along a stretch of road without an emergency lane. I watched children playing kickball get drawn back to classes by a loud ringing bell.

"San Francisco's an expensive place though," I said. "Maybe we'll have more luck down the coast. Heck, most of those places wanted three months' deposit."

"Why we headed south?" Gary asked me.

I replied, "Mexico is south, and I thought we'd get near it." Having thumbed for thirty-five odd miles, we stood before a sign reading, "Palo Alto." We had hiked about seven miles. The roadside gritted, bagging to sharpen the teeth, and got into one's pores. It was easily ninety degrees. I saw a dirt path barely trod and graced by individual flowers, but our heads being exposed were starting to ail. A ride was had from a mid-fortyish woman who told us about the atom splitter that we drove past one full mile. The opposite way was a hillside with perhaps two or one thousand homes, all identical in design. Each block of homes was painted the same color, giving an enormous checkerboard against the mountainside. Quite special, it was. The next ride was from a man driving to the east coast in a Karma Ghia. I got the front seat.

"I'm heading west a few miles, gotta get away. Just broke up with my girlfriend. Too flakey. Tells me she wants to put her IU on her Xmas tree."

"Whadda ya do for a living?" I inquired.

"An artist. I'm a sculptor…say do you pals have any grass? How's about a lid?" he asked.

CHAPTER 2

Too

The night began dropping her scarlet and black veil. We were just opposite a liquor store, and beyond that lay the ocean. We desired to go swimming. Besides, cars at night had less visibility. Gary went to the liquor store, then he held his purchase up to the streetlight, and I went into the store for the exact same purchase. Two girls our age were leaving the beach. I snapped my fingers. They ignored us. Beyond the dunes was two- to three-foot seagrass. From that point, we entered the water and drank, drank from that spot. Yet when we awoke, the moon hung large, and the night air was chilly. I threw the bottle into the sand, making it stand erect, stagger about, and proclaimed, "Holy Mary! California just can't have insects *all* along its coast."

Back at the road, we stood underneath a streetlight for about an hour. We got a ride from a 1940ish sedan in pretty good shape driven by a well-known middle-aged actress Shelly Winters. Having a rear seat allowed me to appreciate the nice touch of an embroidered cord attached to the back of the front seat. I recognized us as from "Boston." In this case, it meant we were "hippies" with drugs being implicit.

"Do any strawberry fields? Damn good shit...ever try purple haze?" Shelly named three or four psychedelics that I've heard of. I tried to ignore their front seat chatter, but she had such assumptions and chatter. My hand entered my jacket's right pocket, and I felt a sugar cube I must have taken from the airplane for nourishment.

"Here," I said, handing the sugar cube over unwrapped.

"You want a sugar cube?" Gary turned around and shot me a mean look.

"No," I said. "I only give these away." And I popped the cube into my mouth.

Somehow, we got this yakker to keep quiet. We drove about ten miles. The silence had me just almost fall asleep.

"I'm going to take you out on the town in Monterrey. First, I have to feed my cat." The sedan veered off to the left, which was unusual for the left side of the road had been all hills, trees, and vegetation. But left, we went onto a half-circle driveway with a one-story house situated on a knoll. The car door was slammed shut, and she went into the house.

"Ned! Do you know who that is?"

"Sure do. What the heck is she going to do?"

"May as well stay inside of something. It is kinda cold out." Gary and this lady kept talking while I tried to sleep it off.

"Just like someplace I've New England," I said to myself, pressing my nose against a fogged-rear window. The road turned right, which was seaside, in the foggy, wet midnight. Fog hovered thickly over a bog. Dew glistened near the car headlights. A shinny wet boulevard of flat cobblestone and a crisis crossed X on a second-story arch greeted tidings.

The nightclub's bouncer was broad smiling and teeth showing out of his red beard, and he said, "I can tell you're not from around here. You two look like nice guys. Tell you what. You can sit out back near the coatroom. I hear the band tonight is pretty good. But no drinking."

I drunkenly reason if I couldn't have another drink, I didn't want to stay there, and I said something nasty to the bouncer. His eyes widened, and he said for us to "get lost!"

The next day was spent on a nearby beach accruing tans, with the only conversation for the day being, "Do you think beards will impede our getting jobs?" It was relaxing, and it was summertime.

Gary got flustered, trying to wrestle me in the ocean—the water acting as a lubricant for my skin.

"It's a night for beers, Frisbees, and T-shirts on girls."

A newlywed couple gave us a ride from their Volkswagen bus, mentioning that the Beach Boys were offering a free concert nearby that night and then saying "goodbye," which was far more preferable to "see you later," which stuck in my craw.

Hours later, I said, "I have stomach pains."

"Want to get something to drink?"

"No tonic or ice cubes such abstinence can have its reward," I said. "Will it be tonight?"

Things became strange. I believed simultaneously, we found ourselves thousands if not hundreds of miles from any home. Two girls strode by our park bench, and Gary tried to follow one girl and to coax her—only he returned nursing a bruise. And I proclaimed, "From Boston." We sat, muted until park lights went out. Blowing hard, the wind blew leaves that even stung the pants cuffs as we tried to sleep.

Before we were touring through this town of Carmel, California, I sized up that there were no house or store street numbers. Falling fast to sleep inside the door well to a rooming house that had heat, we were awoken by the sound of a Bissell sweeping truck. The town being so clean and the day being the flower of the sun, we were glad to be alive. A pancake store opened, and I fell off my stool listening for the first time to the story of Little Black Sambo.

We two decided to stop drinking liquor and to purchase sleeping bags, leaving gear in a tree truck and thumbing twenty miles distance to arrive at an army surplus store rides had told us of. The rough estimate had us with one hundred fifteen dollars. With an hour to kill, we strode into a pool hall which doubles as a tavern—men were drinking beer and shooting pool at eight o'clock in the morning. We had gotten off the beaten path—five- or six-old brick buildings occupying a dust-covered tumble-weeded street. The bar looked like out of a programmed-Western movie. Moxie was added when a very large, beefy handlebar mustache sheriff poised himself by the bar underneath a stuffed antelope head. Two kids, a little bit older than kids, too, were removed from the room to cause consternation at their being present. Light pooled through the swinging saloon doors, reflected off the bar, glittered off the sheriff's badge into the mirror.

The surplus store was a lengthy red brick building. We bought two bright-red sleeping bags, also twenty-five* of three-eighths rope, an Australian jungle hat for me, and a leather sun visor for Gary. We got back to our tree trunk base. It was a fantastic time. Then walking along with a dirt sideways next to the road of this route nine, we were stopped in our tracks by a large white neon sign, saying, "All you can eat 99 cents."

"I'd heard of such culinary delights, you ornery critter," I said to Gary. "But sure, never thought I'd get you one." We left our gear outside the restaurant, not wanting to be a dead giveaway. Devouring meat in particular three portions of it, we then started in on the potatoes and coleslaw. The manager initially gave us a mean look, but we just ignored him. Then he came to our table to say he was losing his shirt on us, that he would give us our money back if we just leave. We ignored him. Then we ate another portion of meat, potatoes, and pickles after the fruit. Gary said to stay away from the carbonated beverages as well as the water, for they would fill us up. My eating brought force to laugh aloud, saying, "You're the savage! El Savage!"

CHAPTER 3

Introduction to Big Sur

A cigar was firmly squared. A late thirties man with busy eyebrows and military black glasses stopped a classy red two-door Chevy. A two-year-old, New York license plates, air-conditioned car was driven by a loose open tie at first glance appearing strong businessman. He had a thick dark mustache, dark skin, and he wore dark sunglasses.

"Hi! Guess you're from New York," Gary said.

"Nope, borrowed the car from a friend. Just kidding. Yeh, I am. Hi. I'm here to shoot a film."

"A movie!" Gary shouted. "Need any extras?"

"It's not that kind of film. It is about Ravi Sahankar and his music. I've been shooting with him in India, and now he wanted to come here and teach some classes."

But I said, "Why shoot a movie about Shankar in California?" Ever step on a rake? How about light a cigarette backward? We drove along the two-lane double yellow center-lined road-going fifty-five, never altering speed. The road was full of curves, and I felt the upholstery, trying to ignore that often, we were changing lanes or coming dangerously close to a guardrail on the right.

Gary in the front seat said to me, "Ned, this is the spot my sister told us to seek out."

"I know," I replied. "What do you think I am stupid? Funny, very funny."

In our direction, the view broke open on the right, exposing the Pacific maybe a thousand feet below and miles of the Pacific Ocean with several mountains showing down the coast. The driver seemed untouched by this experience. Our road rose steadily. To the left was a seven-foot brush all along the way, facing, in turn, large hills that were contagious. The radio being off was nice. Then on the right, the view broke open for an instant more to reveal the panorama below. My stomach got uneasy and twisted. Just as suddenly, only trees and brush were to be seen because we were zipping right along.

"Yep," said the driver from behind his sunglasses. "Seventeen miles of this here road along the edge." What can be described as hairpin turns did not slow the driver who often crossed the double yellow line.

"Hey!" I shouted out.

"You think this is bad?" he said. "You should see what it's like when there's a fog out. Don't sweat it. I've been here about four times."

The sunglasses one pointed out a campsite, which we passed up because we were pointing at a good clip, and we figured there's plenty more space to the preservation territory. Ahead by a few miles, perched curiously to the right side of the road, was a monstrously large weathered wooden cabin with an observation deck on top of it. The driveway was large and circular, gravel, and now I couldn't do justice to the vista.

For below was a whole mountainside, giving away to open air. What must be perhaps a thousand feet below were exceptionally tall trees jutting out of horizontal faces of rock—that was the view from the observation deck. Downstairs literally, that was three footsteps underneath the stony circular driveway, was a handsome wood-encased clothing store. At the rear of this store was a window that defied anyone's nerve to stand beside it and look out. Gary did exactly that while I marveled at the balls it must take to create such a window.

"Care to share a cup of coffee," I asked a lovely salesgirl in an attractive sari. She looked at me, who was standing before a mirror wearing a Mexican wedding shirt and looking not bad with a few days beard.

"I doubt you could afford it," she said. "Besides, I think I have a date with a waiter who is upstairs there."

Atop the rood was penned in by two-inch railings, hardly substantial footing. There stood two or three couples admiring the view. I studied the view for some time and estimated that it would take some two weeks to reach the coast below and to return.

"Impressive. Isn't it?" spoke a man standing next to me. Balboa had seen the Pacific for the first time right down there below.

"Did he make it?" I asked, curious if it was in the direction I had thought of.

"Yeah. He made it. He had to go thirty miles north before he first got to the Pacific. He gave the view below a sweep of his hand, saying that after a month of excursion down there, he found he had made only five miles. It's some jungle."

"Jungle?"

"Well, call it what you will, but just look at it." I let his words take over and see that one who sought out the Pacific below was risking the hair one's back.

Looking about for Brute Force Sandford, I saw him down in the parking lot, where he was avidly pumping the hands of the movie producer and Ringo Starr and Ravi Shankar. As I approached, it was revealed that Mr. Starr apparently was considerably tied to being recognized at first sight. I stayed away from them while a man observing, in turn, my watching them gave me a unique look. That man was Ravi Shankar. Gary ran up to me and bumped into me.

"You see who that was!" Gary said. I was snakes hit in my interrupted thoughts and refused to go over and greet those people just for a handshake.

Prices in the Phoenix restaurant were expensive. Just where were we? There was the provision store six miles back down the road or north next to the camping area, a lone gas station south eight miles. It was saved for the reservation territory itself, Big Sur. Gear in hand, jackets over the shoulder, we walked with backs to any traffic after eight cars passed, each flashing us the peace sign.

"We could be anywhere in the world," Gary said, patting my shoulder. "And this would be what I'd seek."

Our eyes noticed the cap sight had numerous tents and cleared open spaces for campfires. Everyone was gone. We could have a field day or recognize all are about the world enjoying the world that we share. At the provision store, I put a package of cheese into my midriff then pulled my shirt over. There was a package of bologna inside my hat. Gary likewise put some bologna into his midriff. We paid for some small cheese and crackers, then at the doorway, we stole two kites and balls of string.

"You must be from not around here," the clerk said. I noticed his beard. We exited without another word, walked twenty paces, and sat on the ground.

In an unsurprising moment, the state trooper pulled his cruiser right next to my head.

The California state trooper took my ID. "Let me tell you something," he stated, "I can tell you that the two of you think because of where you're from, you are something special. Well, I'll tell you that I could put you in jail for days without you seeing a judge, and in jail, they'd eat you alive." With a twist, he said, "See you two later, sure as hell." We were in no mood to get into more trouble and moved away from the campsite, walking six or so miles back to the Phoenix.

Next came a full-sized bridge spanning what might be another thousand feet…the Pfeiffer bridge. I stopped for a second on my stepping, and I felt tired. Even my legs were tired when Gary said, "My legs are shot. Let's thumbs from here." I looked at the span and the chasm below and imagined getting halfway across and freezing in step.

"Look, pal," I said to Gary, "whatever happens next depends on who picks us up. Now let's get our gear together and get at least a ride if not a dinner out of the next car."

Three women in a Ford Mustang came along, and they stopped for us at the start of this remarkable bridge.

"Linda, you get out [of the front seat], and, Pam, you get out of the car and let that man thereby the rear of the car get into the back seat with you. You sit on his lap. And, Linda, you get in on this one's lap half on the console and half on his lap, just so the other one doesn't get to thinking he's better looking than the other," said one of the three women.

I tried to get the girl in the rear seat interested me. It was sweet having her on my lap, as attractive as Pamela as I had ever met. But she stayed averted from me. I breathed a shot of warm air into her ear.

"Phew, you smell! Buy a toothbrush or at least use one."

I held my palm up to my nose and blow. Nothing, not even in the front seat. And three times, I couldn't stand looking out because I was scared. I thought of the poor motorcyclist who might have to cross. Once across, in one long minute, we were let out! Pointing to a very steep hill across the road, I suggested that we climb it to watch the sunset.

I ignored that Gary was passing on a dare when he said, "Look at the size of that. It's almost a mountain."

And I said, "Who cares."

CHAPTER 4

Three Thousand Dollars' Worth of Drugs

Big Sur was just too beautiful for words, so I wrote helplessly save for his grace.

Ahhhh, fifty in, it was noticeable that we were not pacing on the pavement. Soft sod was underfoot, and all it was pretty and quiet. The surroundings were tan and beige and dark green, but something was just not quite right. Trees were spaced evenly apart about fifty feet. But they were standing in the shade. In fact, there was shade everywhere, and arching heads revealed two-hundred-foot trees with branches starting at the eight-foot line.

"Will you just look at those trees, Gary."

"We come all the way to California, and you want me to look at trees!"

One looked closely at the other, both thinking face-to-face, *Is there something wrong with him?*

A slightly showing path led us into the forest. Pine needles were cover the rising way. A felled tree blocked our path, and it had such girl that this fourth son failed to explain it. We used all four limbs to crawl over the tree.

After many yards, the timberline broke open. Even green and majestically broad, the timberline provided a contrast to the rock ledged multiple boulder mountainside. Limitless, the Pacific stood. The mountain was uneven, but to the far left, there was the hint of a ten-inch-wide trail. The occasional white flower growing straight up was noteworthy. I climbed at a pace faster than Gary, and soon I had some distance between us. I climbed faster, thinking crazily, *Will it be daylight on the other side of the hill?*

When I stopped a second time to rest, I saw I was too far up the trail to receive any help. The night was inching its way up the mountain. Single twisting, purling streams of smoke rose from the campfires beneath the timberline—their color different from the night. I sized up the situation. I was on my belly, using my right arm and left leg, left arm and right leg alternatingly. Paul, Luke, and John Kenneth Galbraith would never want to view the open terrain falling away from the hillside.

The summit was another hundred feet away. Darkness was imminent.

"Hey! Don't come up anymore!" I shouted. Gary continued climbing up the trail, stopped, then stayed in place. I got on my backside then I slid in patches down the trail. Together, we stood up to stretch our spines. That I might have been up that trail flickering pebbles for my prayer beads was of little solace. Night fell. Gary said, "People back at the provision store warned me about boars in these hills."

There wasn't a moon. Visibility was narrowed by cloud cover, but after we got into sleeping bags, I could see Gary hunched out like a ghostly wraith silhouetted against the mountainside—to walk out and disappear only to return with rocks that he put at the foot of the sleeping bag to prevent sliding. I vowed to stay put, not trusting to earlier seen ten-foot drops. Shivering throughout the night despite putting on all our clothing, we peeked out of our bags and saw that we really were sliding down the slope.

"It's a good thing 'bout bulls being color blind," Gary said as he noticed we boys were surrounded by a small herd of cows and that our sleeping bags were each a bright red.

"What makes you so sure about that?"

But when they turned around and saw the lone fig tree at the summit just somehow indicates to be content with heading for flatter land. We noticed the beauty of the extensive Pacific. The ranges we're the start of where the Santa Lucia, the timberline at hand, and the occasional stark white flower.

Good morning—it was so special for Gary and me to look out from a mountainside to observe below us a vista of miles of white clouds, a spectacle of bulbous clouds stretching miles with the blue Pacific Ocean hundreds of feet below this remarkable view.

We traced our tracks, again going over that ever so large and wide fallen tree, and when we arrived at the road with the pavement under your feet, I turned and stepped back onto the forest bed to remind myself of that feeling on my feet. I understood there are many miles of pavement to come. We walked to our left from the mountain. It was toward Los Angeles, and it was downhill for a good stretch until we arrived at a spot where a car might stop and not risk being rear-ended. We thumbed for three hours until a green car twenty yards ahead of our thumbs let out two girls attired like us—their passing car giving was the peace sign. The next automobile stopped for these two denim-clad girls. They smiled at us and flashed us the peace sign. I screamed obscenities about women in the wilderness.

I noticed the view. There was a guardrail to our left besides us, and behind that stood orange trees. We stuffed our bags with oranges. In ninety minutes, there was a ride from a blue Volkswagen bug driven along by a blue-eyed bearded twenty-something-year-old. We all chatted along gaily for twenty minutes at least. Then the land became level, the forest disappeared, and the ocean shored up to our right.

"We'll get out here," I spoke with gusto.

"You'd better not swim there."

"Why? Sharks?" I said while running alongside the Volkswagen and trying to close the passenger door.

"No," the man said, engine whining away. "The barracuda scare them away."

"Boy!" Gary said, upchucking, bending over to hold both knees. "I've been holding that down in my stomach since we first got in the car."

I laughed fully, holding onto my own shaking stomach and knees.

As I said, the land was now flat, and cross the double lane road was a street with a few housed homes. And there stood a single sign, which I pondered for many a minute. A car came to the stop sign.

"Hey, mister, can you give us a lift?" Gary said.

"Sorry," the adult man from sitting ever so comfortable in his car said. "I'm only going a few miles that way."

We two-lane hikers stared blankly at one another. Then one of us two asked, "Can you let us know where we can go swimming. And wash?"

"Well, you sure can't go there." He pointed to the sea. "Yah, you can go back into Big Sur a few miles and make your first left, which is a single-lane dirt path. I doubt you'll make it before sunset," he said, keeping something to himself. "You'll see why."

And after receiving another ride from a haired, bearded man driving a bug, we saw it was a left turn all right. The road consisted of two adjacent paths running together—a well-worn two-path lane. This entire roadway under a close to fully extending canopy of trees and brush blazed beside nearby thick greenery growth. The roads went on and on as we paced it out underneath this "tube" while on the brown and gray floor. Various motley colors flickered through. Time after time, one of us would say, "It can't be much further."

The forest broke open to reveal a half moon of sandy soil flanked by palm trees. The smell of sea air was good and reinvigorated. Nothing would surprise. Darkening the scene were two men, both sizeable one, red-haired, and bearded—both toting rifles.

"Pay two dollars," the reddish one said, pointing his rifle near to my stomach.

"You didn't even say please," said Gary, standing still in the cleaning. It was half an hour until total darkness. From the beach area, an approaching whining sound was heard…a Volkswagen! Gary folded his arms together and stood in the middle of the start to this dirt road as we looked inside the bug at three long-haired miles inside, outside of it two made older than Gary and I who footed the rear bumper.

"You didn't have to do that," said the driver, who was spectacled, granny glasses perched on the nose, shoulder-length straight brown hair. "Climb up on the rear bumper with the other guys."

We started going, but one of the long hairs on the rear bumper pounded on the VW roof when a package dropped off the luggage rack. And although the car was moving, one guy jumped off and retrieved the small suitcase.

"Boy!" he said to his companion while he was back onto the bumper. "Can you imagine all your drugs being lost? That's worth at least three thousand dollars." Gary winked at me.

We were dropped off (the VW holding true) back where the land leveled out, which was nice as it is a number, say six of miles from the earthen roadway. We noticed a house across the road, and in the driveway was a black Labrador with her litter. We left the road and went playing with the puppies.

"There's too young to take," Gary said, canceling my notion of using the dogs as a device to get us a lift.

It was night, and with little else to do, we unfurled our sleeping bags and lay down in the road's emergency lane. Putting the rest of our gear under our ears, we closed our eyes. A car approaching from Big Sur was in earshot, and I belly laughed. Sure enough, a Volkswagen minivan stopped for us. And I ran for the back seat while Gary entered the front with a most pallid "hello."

CHAPTER 5

Thermometer Outside

Military eyeglasses, hard-eyed, balding, a forty-year-old man with bushy eyebrows brought his four-door Chevy with radial tires to a stop. Letting us into the car, he said, "The way two look, your best bet is to take the highway situated inland." Gary and I disagreed, as we were just going to go coast way. "But," he continued, "the coast way is pretty. That's for sure. But not too many people going that way won't be inclined to pick up *two* people who look like you. Believe me. I'm from around here, and nothing personal, but people like you are coming through every day. The highway is full of people going to Los Angeles, and you'll have no problem."

After his third innuendo concerning our appearance, I gushed out, "All right! All right, we'll take the inland road!" None the worse for wear, we had a good stretch of roadway traveling with this man. We were let out in the night. We walked with our backs to the traffic. Forty full laden trucks *zoom-push-swoosh* left our feet from the air push—being two or three feet along. Hands did not sweat because of the grit. Dirt continued to sharpen our teeth. It was cold, but sweat formed on our foreheads. My head, aggravated, suffered a bit of an implosion each time a diesel roars by and poured out more noxious fumes. A quarter of a mile walk and Gary shouted, "Hey! Let's go inland."

His words did not register with me, and I continued to walk. Gary trotted over to me, spun me around, saw a painful look on my face, and he brotherly put his arms arm around me and led me off from the highway. In a ditch, just a few feet from the road, only a bit of them way past the trucks, the night was passed.

Neither of us understood that it was 175 miles to Los Angeles. What is known was how to be good days thumbing would get us there. This morning started with a two-mile walk before a car even slowed down. That car stopped thirty yards from us and backed up. Running ahead to the driver, I asked if he minds that I sleep in the back seat. I'd slept by the road and didn't sleep all night.

"No problem," said the man, who was three to four times older than me. He announced, "I'd be glad to help you out. Which ways did you come in? Ha, ha, ha. Sorry, just an old joke, older than the two of you together. I'm offering a ride to Bakersfield, not directly to LA but in that general direction." I fell asleep right away inside the auto. Later feeling the road surface changed, I thought that we've opted for a mistake.

"How much further you going?"

No reply.

Inside of ten miles, we got our reply to the question, "Is this the way we're going?" Out of the car, where are we? In a desert. In such as this, I remembered begging for relief from junior high, where I'd flutter through a book, the pages yes or no capturing my interest to hook me.

The road consisted of two straight hot and flat lanes. No fauna saved the scant rawboned tree. For hours, we'd walk mile upon mile, step after step. Amusing myself, I stepped onto the dirt, making tiny bombs. That was until Gary stop me from wasting energy. We rested and then continued walking. We each took turns at keeping our spirits up. We shoved one another, made some silly pun, or just laugh at our situation.

"Ten bucks says it is going to be ten degrees hotter inside another hour."

"You know it's going to be twelve degrees hotter." We had walked about ten miles, arriving at a lone bright red-colored house. The place doubled as a real estate local and supermarket, albeit small.

"Got an appointment with the devil down the road in an hour. Let's kill some time," Gay said, faking me out by walking past the store, and he pulled a tug on my elbow. The fairly large grocery had a plentiful array of foods in this year of our Lord, 1968. The middle-aged white smocked-glasses clerk greeted us with a large smile but soon had his arms folded across his chest.

"Can I help you?" the proprietor asked.

"Yeah," I replied. "I want two banana popsicles please, two of 'em," We paid for some crackers (small), the popsicles. And the man refused to let me back in the store.

"You've had your chance," the man in the white smock countered, wearing a forced smile. Outside, we walked in that rare thing, an emergency lane. The sun beat down on our backs and neck. I contemplated the possible pranks of nature (and if the sun would be pulsating every single day for this excursion of ours). I wondered out loud if God would punish us for stealing a second time. Gary told me to "shut up."

"Maybe we'll have to walk to LA," I spoke.

The next morning, as light broke, I observed a small bird hop out of a hollow log, yawn, and dart away into the day. The sun started to wear on our necks. We arrived at a small store with wooden stairs that we sat upon. Along came the owner, saying, "Looks like you two could use a day-old doughnut and a free cup of coffee as well."

Baby, once inside the place, I sat up on one counter, trying to look cool. All too soon, I felt myself falling backward into the waitress pit. The owner rushed out of the kitchen baking area to tell us, "Get the hell out of here, you!"

Outside the store, just as we walked down its steps, the morning newspaper was delivered. I picked it up, ignored the headline, and discarded it unread. Then we walked and thumbed a few miles, mostly walking.

We arrived at a vantage point high above the city of Bakersfield. There were numerous huge boulders on both sides of this highway.

We looked down at the city of Bakersfield. The city was surrounded by desert and would hardly exist save many a hard man's toils. I reasoned from my spying position that Bakersfield must be an important city. Down in there, in Bakersfield, I observed, were countless autos and trucks.

There was another lift from a man in his late twenties. "Man, did it rain when you were in Big Sur?"

"Well, you'd know if it had rained. The hills and mountains of Big Sur form raging streams that consume trees, rocks, sweep up anything in its developing current."

Rides from longhairs were friend, informative. Whatever the subject, one would just hone in on it.

"Don't hike on any California state highways."

"But we haven't any pot."

"That doesn't matter, man. The regular troopers are really tough, stringent."

"Whadahyamean?"

"They don't give anyone a break. It's two hundred jail or both. So don't step out. The regular roads are cool though." The cars we were let into just said, "Go inland. Slower but safer all around."

Left outside at an off-ramp, we walked up the ramp to the highway. At one coalescing point, Gary and I met two girls from South Dakota who told us they had been raped. The girls were large, larger than either of us boys and heavy. I reasoned they were in a situation like us, just broke and had to thumb plain out of bread. One of the girls said they also had been robbed. The other one, Susan, had dirty eyes, straight black hair, and her eyes danced. She drew out of Jew's harp and began to play. We guys shuffled our feet, looking at one another, doubting that they had been raped.

"You got any green to spend?" the stockiest one inquired, laughing. She gyrated her pelvis, and our eyes widened at how large she really is. The subject was understood. Looking at the ground, still shuffling out feet, we said, "No, thanks." When we were separated by a few steps, the girls talked to one another. They grinned and laughed.

"Sure?"

"Un hngh."

When one hitchhiked, one better be prepared—prepared to defend your life. Jilt the jugular. Prep with a hatpin, a knife, a single edge razor. It was not usual to get picked up by a total stranger who said he's "ripe enough to use a can of worms." A hint, say it's against your religion. They will go another ten miles before a single question comes up.

The girls got picked up first. Fifty minutes later, we had a lift. The ride was from a self-stated forty-three-year-old man with his eleven-year-old son. This ride was from a pickup truck with a camper attached to its rear. The air conditioner was turned off, which was a good idea as the thermometer of the passengers' sides reads, "One hundred twenty-two degrees."

"But," the man said to his son, "that is because the thermometer is directly in the sun. Nobody would be nuts enough to stand out in the sun anyway."

"Did you stand out in it?" the boy asked.

Later, on foot approaching the highway, we eyed a California state highway police car that drove slowly by us. The trooper pointed to a sign saying, "No hitchhiking." He pointed for us to thumb from the start of the ramp to reach the highway. The second car passing stopped very quickly, and the driver hastily warned, "I know you must be from out of state because if you get caught thumbing on highways in this state, you get fined seventy-five dollars and have to spend five days in the fridge." He drove fiendishly over ninety but did not reach the trooper then out at an on-ramp.

The next car stopped, and the driver spoke immediately, "Do you guys know that…wait a minute, this isn't emergent, is it? Because you sure don't know that if you get caught, you'll be given no except and put in jail for a week."

"That so."

That ride was for fifteen miles.

The next ride was from a pickup truck. We were handed a ride from the rear. We spread out our sleeping bags, and each used a side of the rear compartment as a railing to rest our arms. We went through a gap in the Tehachapi mountains. What we have you see was a comfortable rise as if in captain's chairs—the sun strong coloring, swirling wind. It came in refreshing, sweat-assisted gusts. We climbed in the truck from the flat sand-colored clear-cut desert. I rose to put my hands atop the cab, and the wind sliced my hat off, but I caught it with a quick hand. I laughed with gusto for a while, and the two men in the front seat turned around and smiled at me. At seventy miles an hour, I read through a gap, through a now timberline as we climbed. The view of this exciting country was (hands up) westerly beautiful. Gary yelled in glee. The drop to my right at one point had to be of four hundred feet, nothingness by the edge to the road going in our direction. For thirty seconds, I stopped to admire the rear window reflection of my grin.

There were still no clouds. We were not thirsty. The shining green truck rolled slowly away with its plates conspicuous while fitting into the surroundings. We started walking again, first with confidence then with increasing moodiness.

"Damn it all, we won't be getting to Los Angeles today," I snarled. Back at me but kindly, Gary said, "We can spend the night here—only we go away from the truck noise." The incessant roar of these semis!

CHAPTER 6

He's a Dead Head

The city was thirty miles away. Deciding jail or no jail, we had to get to Los Angeles today. We took on a ramp for the highway and walked in the emergency lane with our backs to traffic. Although no car was going to stop at sunrise, we were wary of any approaching vehicle. The occasional truck roared by beating daylight traffic. After a pair of miles, we arrived at a gas station and had peanut butter and cheese crackers, washing it down with a coke. After an hour, we had been loitering so long that a gas station attendant told us to "get lost!" In an hour, we had a lift.

A clean, sharp white Chevy with its four windows rolled down went from the passing lane to a dead stop in but yards. The man was wearing a white T-shirt, sleeves rolled up with his right arm rolled up to the shoulder with a pack of camels inside the T-shirt.

"Going to Los Angeles" was all he said at first. I was impressed with the man, and no matter how secure, casual, rugged with muscles bulging he was, I was not about to interrupt a seventy-mile-and-hour ride with fat-chewing questions.

Gary, in the back seat, said, "Just look at this land." And we rode comfortably, admiring the vast San Fernando valley fertile green acreage—mile upon mile.

"I'm about to hitch across the country. Just came in from San Francisco," I said. The driver nodded his short-cropped head, saying,

"I'm returning from Chicago. I'm a troubleshooter for the 'X' union. I go wherever needed to stir things up…start a fight…whatever."

LA, we approached. The highway expanded to seven lanes in our direction. The bumper hedging was incredible! Billboards sprang into view. One product was especially noticed funerals, inexpensive, nice funerals. This was advertised on busses, billboards, park benches, and later on, television.

Ours was a nice chitchat about sports, the weather. The obvious encroaching manifest of the city was ignored. At the red traffic light, we hopped out. He closed the passenger's door from the inside, reciting, "Goodbye. Radio says it's going to rain."

"That so. Well, thank you very much."

"De nada."

Off to the right, I noticed a weather-beaten shack, and my nostrils flared from the nearby sea.

"Venice, just down the road," a few street people said. "Folks there will put you up, we bet." We felt comfortable about to fit in. They continued, saying, "There are people like you there."

"That good or bad," I said to Gary. We walked away. Then right beside a sign reading Venice, a police car pulled over for us. We were investigated, and the policeman of two who was interviewing me said, "What do they call you, Ned?" I realized my name was on the photostat page of identification. The Pacific was our backdrop.

"Did he keep that page separated from his wallet?" the other policeman asked.

"Yeah, he has brains," my policeman replied. The police were friendly to the point they said, "Look, take care of each other. California has some strange people, and the city has some weirdos. Watch out for yourselves."

I donned my sunglasses for the first time on the trip as it was beginning to rain. The ocean invited us to visit the water edge. For a half a mile, we walked along the edge, and the wind blew clusters of sand against our faces. I was glad I was wearing sunglasses. The dark green sea was rolling a small storm spit cups of water onto us. We could drink the air, mon frer. As the wind ceased, we could see through the rain that up the coast were modern-designed houses at the edge of a row of cliffs.

"Homesite can improve one's thinking," Gary said, impressing me, his partner. So I patted Gary on the shoulder.

"What is it?"

"Look."

"What?"

"Look at me."

"Yeah, what?"

"I'm smiling, idiot." We walked away from the shore's edge, keeping a separate pace.

Near a series of old brick buildings and between the sea and the building was a well. A full-bearded, long-haired man sat in a lotus position beside the well, and he clutched a brown paper bag with probably a bottle of wine inside.

"Isn't the day too difficult to use drugs?" I asked.

On a straight street stood six buildings three to a side. In between these three-story buildings, away from both the stinging sand and the rain, some seven or eight people were playing in a game of tag. They noticed us approaching, and they huddled, pointing at us, two approaching strangers. Taller than the others, a man with a beard, a little bit similar to a Van Dyke, only more like a Ho Chin Minn beard, walked with a female blond toward us.

He asked us if we have any acid.

"Un un," I replied.

"Wanna buy some?" he questioned me. Few sundry questions were for the asking by these two young twenties folk, both in need of a shower in the immediacy of the situation. She asked me for a dollar. He was staring at us with very intense eyes. Wide-open, his eyes had a controlling look. I sized the guy up as being the leader of the group, still huddled, still pointing. Quickly, he had a smile for Gary. Gary was likewise interested in the hombre.

I stepped back one and a half paces. I saw this anything, but a kid had an interest in Gary. This man viewed me cautiously, then he turned back to talking with Gary, saying he got a place away from the storm. He had some booze, women, and food for us. Gary, for me, was obviously pleased that he had found a place for us to hang our hats.

Yet clearly, my inner voice was speaking to me, saying, "Get the hell out of here. Now. Pronto!" Gary's eyes were soft green and reciprocal. His face showed surprise at my saying for us to leave, but he agreed, saying, "I don't understand or agree with you, but all right, I respect your opinion."

Six years after this trip, I was reading in my basement at eleven thirty at night, and I came across a passage from the book *Helter Skelter*. There was a passage that ruined my blood cold. Vincent Bugliosi stated in his book that during the summer of 1968, Charles Manson and his cohorts hung out and sold acid from dilapidated apartment buildings in Venice, California. It was the group that Gary and I walked away from.

Allow me to continue, hitchhiking down California's highways, three times a yellow school bus drove past us with a long-haired freak at the wheel. The third time this man appeared to be heading into turning the bus over to the side of the road for we two hitchhikers. I gave him the finger. Manson showed a smirk and a smile that he was cruising while I was thumbing.

CHAPTER 7

Pancake Makeup

Beside the sidewalk was car after car. It was an unlimited view. I saw a street sign, a typical one. It read, "Los Angeles." At first, we hitchhiked from the sidewalk, ridiculous with the proximal street full of parked cars. Gary rested nearby then took to the road, saying, "Let me try."

"Ned! Come on, Ned, a ride!" I turned to see. Steps behind me, Gary was getting it to a candy red apple dual seater convertible. The sports car was driven by a red-haired young gal wearing bulbous red sunglasses. She wore a Celtic green miniskirt going up to her great gams and a pink sweater that was mohair and demarcated her assets. Tossing my bag onto Gary's lap, I climbed on to the rear of the car and held on to the underneath with one hand.

Occasionally, words flew up through the wind of this sports auto: "Big Sur," "really," and "Boston." I looked around, then I studied my right hand. I laughed at the world.

The woman drove exceptionally poorly—maybe she was trying to impress us. Gary gave me a wary look. A right then left and a sign resting in some greenery read, "Beverly Hills," then a right and sharp left into a single lane driveway. I left the car behind and started walking for the main road. Gary said, "Hey! Where are you going?" He indicated for me to follow both of them up a flight of stairs as he smiled at her derriere. At the door, she opened it without extricating any keys.

"You don't lock your door?" I asked.

"No," she said. "Why should I?"

Entering, we were inside a one-room flat with a small kitchen and bathroom. There were two couches that folded down to beds, a table, two chairs, a television, and filling the room were numerous throw pillows. There was a thick shag carpet. "Erin is my name," she explained. "I'm twenty-three. You can have my apartment for a week while I'm away with my boyfriend at Big Sur. You were where I am going to be going with my boyfriend. I was here where you're going to be." Erin oozes out, "And we're all from within fourteen miles of the oceans, only they're opposites."

Inside a scant ten minutes, there was a tall thin man who came in and gave half a bow to three others. He and Erin rushed out the door after grabbing a backpack and sleeping bag, and leaving through the door, Erin said, "There's a box on the table. If you can open it, you're in for a surprise. And there's no key, so don't you lock the door." She closed the door.

Scouting the neighborhood, locating a diner, we each digested a meal that greatly reduced some real stomach pains. Erin was loaning us her flat, so we won't be eating any of her food. At a guarded large market, we purchased a huge box of pancake mix that should have been labeled "Army size."

Nine miles from a beach in Santa Monica. A home. We fought the couch/mattress for an hour, switched beds, and then slept until noon the following day—twelve hours. Stepping outside, I noticed our spritely pace and how rested we were, mentioning this to Gary (a.k.a. Brute Force).

"What the hell are you? English major?" Gary dared defiantly. I grew defensive, drew inward, looked at my feet while thinking, *It hurts to think you're too outspoken with your best friend.* We continued our walk then thumbed to Santa Monica. It was the beach we headed for—the football and T-shirts off showed us for what we were page white and new to the era. Children tossed in the ocean's two-foot waves, rafts going up and reflecting the sunrays sharply in the water. We caught rays sunning on the beach.

All the beach bathing made for a little world and thought bliss, up wrestling in the water every hour. There were plenty of girls. It was a weekday. Few folks our age were in the area—in fact, just one way down the beach. Eventually, two blond girls walked past us. Jumping up, I followed them and came to a beach house where there was a volleyball game going on. Not familiar with the sports and they kept the ball away from me, I headed down and dejected. White as sin, I walked away from the gathering. I walked toward the Santa Monica pier, and by the time I reached it, an older teenage-year-old girl, Marie, had followed me to speak, "Looks like they were blackballing you. I saw you at the volleyball game."

Conversation became close and friendly—familiar chat. At the arcade on the pier, I bought Marie an orange drink, and when she finished it, it went in for a kiss. She made a soft slap on my face and said, "Hell no, my old man is a hundred yards away, and I bet you he's watching me right now. Better off him seeing me reject you. Maybe after dinner is possible, very possible, cutie." I watched her leave, going up the beach to the volleyball net, and I thought of her phrase, "My old man." Once she was gone, I ran for the restaurant that was on the arcade that was advertising "Hamburgers twenty-nine cents." Ordering "one," I saw boxes of cracker jacks and salivate. It had been about a meal every day now.

Nobody could really say what I looked like when the waitress saw me ordering my hamburger. I was in cutoff Levis, a goatee, and was noticeably missing a tan. The waitress seemed to look away when she handed over my meal despite that a sign directly before me read, "Pay when served!" Gobs of relish and ketchup. Sauntering past the cash register, I exited to the pier, feeling guilty as I opened the door. I ran along the pier and hopped onto the beach, where I ran for three-quarters of a mile. As I stopped, I experienced the worst stomach pains on the trip to date. The pains were really something. This one really tried my constitution. The longer than thirty-second seizures were really something.

Inside a larger than life cement tunnel, one passed from the roadway through to the beach. No graffiti or urine was present.

We browsed through the mall by the sea in Santa Monica—aft shows, musicians, flowers, craftspeople. We people watched on this ninety-degree day. One day, we took time to appreciate a nearby park and Jimmy Webb's poem/song MacArthur's park and the elderly sitting and telling life's recitable story. One evening, we watched a pair of fencing then cut through a backyard to see a man and woman making love through an open curtainless window. The woman was amused. Gary put his arms on the windowsill until the male drew the curtains with a huff. Trotting through the same yard, we reached a shed, and inside, I found in the dark a horse blanket…red, black, white, and gold. This I'd use for a pillow.

One day, on our Santa Monica freeway, I printed wet footprints leading from the fountain where we had been resting our paws to the mall space. I heard someone shouted, "Ned!" I paid no attention to this. "Ned" was shouted, and home I continued my pace, knowing who I was. Then loudly, "McFadden!" was yelled, and I turned to see from the sparse crowd my neighborhood druggist. He was dumbfounded that we were meeting. The man was on his honeymoon. I said, "We're thumbing across the country." The druggist got a huge hoot and hollered out of this.

We headed back for the apartment. To the outside of parked cars, we thumbed. Each of us wore a smile. As I rested my eyes, a black two-door Chevy passed with its passenger hanging out from the open window giving me the double-barreled fickle finger of fate. It was the same guy who in Massachusetts told me, "See you in California." Mark Loud. Gary openmouthed looked at me. We continued to walk backward and thumbed, ignoring warnings from several people not to thumb here, these cars, again and again as they flashed the peace sign. Occasionally, they, too, were given the finger or two.

How I hadn't taken notice…hubcaps! Hubcaps, everything one could imagine on parked cars, the passing cars, the cars down the street, cars on sale, and cars on trailers with sixteen times what hubs! Metallic blue, fire engine red, yellow school bus, chrome, silver

GARY AND DUKE

plated, aluminum sprockets—Los Angeles treasures. Five full minutes of this folly and I saw Bentley's hubcaps. My face surrendered a childlike look, and the Bentley noticed me, pulled over, and stopped. I stood flat-footed, which left Gary to only go for the front seat. Soon, I was noticing my dusty boot brush against the automobiles brushed mahogany bar.

As I listened to the two in the front seat talk about "Boston," I saw my reflection in the mirror and listened to the man. He was impressed with Gary and would be happy to show both of us a tour of Beverly Hills. It was his sincere pleasure to help out we two who have traveled to see the lights. He was about fifty (couldn't really tell), silver hair, glasses of brown with a silver streak. Mansion viewing and city history, we toured with this man. This was a treat that pleased me.

Manicured laws separated residences up and down their hills. We went up a hill, a lengthy driveway abutting thick woods and brush. Well removed from the other houses and the street, this one house prompted me to speak.

"It wouldn't hurt for the person who lives up here to have his own fire engine," I specified.

"Here now, he's already got a fire engine. He's my brother. When he moved here a couple of years ago, a mutual friend kidded him about exactly that his state is so far from home that there won't be anything left by the time the fire department gets there. I mean here. So he went out and bought one…a fire engine. Right away. He had some girls in bathing suits hired to ride to with the fireman's hat. And he drove it all over and right through downtown with the sirens on full. He got stopped and got ticketed. But he didn't care."

"I thought Los Angeles didn't have a center," I spoke.

"Oh, there's pretty much a center, the business part of Burbank at least."

Upon a hilltop, there was a stucco mansion. A pavement ending calmed sac circular driveways end—a green XKE, two matching Ford Mustang, a red European sports car, and a medium-size fire engine.

"See those hills over there. You're really lucky to see them because usually there's smog. That's Frank Sinatra's mother's home next door." He explained that he and his brother were then coinvestors of the Cathy Doll, a doll that made millions and sold millions. "It was a device, the device to say a multiple of things that was the real problem. Edison worked on it, too, but he only got it to say one thing." He had made it, sold it, and on it made millions in royalties.

We saw no pedestrians, no lady out with a stroller. Still, it was a mental boost even if there were trees, hills, and manicured lawns. The man offered to let us stay at his garage-housing quarters with a few graduate students.

"We already have a place. Thanks."

Next, back at Erin's place, we ate several helpings of pancakes. Gary went to watch television, but this struck me as awful.

I said, "I'm going out for some." Gary ignored my "goodbye."

I decided to solo it to Santa Monica, to run a couple of miles along the beachside. I said hello to several people. That was until a dark hombre wearing sunglasses a half hour after sunset stared at me. Hitchhiking back at a stoplight, I saw that there was no traffic, and my seeing a woman stopped in the opposite direction, I said, "If I cross the road and thumb that way, will you give me a lift?"

The woman laughed gustily. "Mmmmmm," she roared. "You are a doll, hot pants. What are you some dude hitchhiking across the country during the summer off from college and getting your rocks off?"

"Hungh," I said. "I guess so," I said with the wool off my neck, feeling my neck looking for a line.

She was still laughing delightedly, confident, secured, and amused.

"Oh, honey, don't mind me laughing. It's not what you think. I'm a lot older than you. I'm forty-two. I'm only laughing because I like the idea. I think you must have a lot of balls. And I don't mean that as a joke, but to thumb around a city like this is dangerous a lot, believe me. And I'm telling you respectfully, to appreciate that I'm older than you, giving it to you straight. You've gotta have something on the ball just to be doing what you are doing. You're not alone, are you?"

"Uh, Uh."

"I didn't think so. You look too smart for that. You're doing the right thing...take my word for it. Take it...Get it all while you can, yes, sir."

I was sizing up her car, her features, and her character. It was pleasing how this gal just opened up her pocketbook and spoke. It crossed my mind that I didn't want sex with her because I like her.

"But I got horny one night," she said, "and got knocked up by seventeen. Three by twenty-five. Shit, I don't even know if I like them any more now that they've grown up the oldest one that is." She paused reflectively. "I still do like Chrissy. She's the youngest one. That's who I was going to get to pick up when you spoke to me and got so funny. You're some character, mister. Had any lately?" I cleared my throat, and the woman laughed a gusty laugh, too gusty for my taste. "Ha, ha, ha. Don't worry, babe, I'm not getting you up for a lay. I'm cesarean. You wouldn't be interested in an old babe like me after the boots I've been wearing all these years."

I found my manhood dared, but she was so damn bland with her sex chat. At a red light, I opened the passenger's door and got out.

She said, "Oh, really, it's not you...it's just that men are such bastards, and I am having problems with the father, and he's coming over tonight, and I don't want him and..." The next ride home, I acted as if I was mute.

The calendar passed another day after we again had slept for twelve hours. My complaining about our lack of exercise got Gary toping me against the coach, twice in two-ear splitting times. The times arrived when it was too late for the beach and too early to go out. We had been listening to the stereo. Gary warned me not to play the Beatles "because I'm sick of 'em." I went over to the box, the hand-carved box that our hostess Erin had spoken of. I fingered at it, but it didn't open. I refused to let Gary try, and while reflecting on my reflection, it split open, revealing a plastic bag with some grasslike material inside. A small pipe too was taken from this box.

From a memorable sight from a scene at a party I had once attended, I went to the kitchen and procured some aluminum foil, which I put small holes in and pressed the bowl shape into the pipe bowl. I some a pipeful. Gary came over for some, which got agreed to.

"I've heard you can't get high the first time," Gary said bullishly.

"Nonsense!" I snapped back—another pipe. Halfway through another pipe and I started to notice the music more perceptively. I began getting a step higher. The music on the record player was by the group Crème, "I Feel Free." I was drawn above little aches, little problems. I thought it wasn't any chemical making me tick differently, just my thinking. Next, I was lost and decided to close my eyes. Colorful images entered my mind with the aid of music. I was in a lotus position with my eye closed. I felt I was levitating. I really did. I gazed out of my eyes at Gary, who was cute as he stared down the pipe bowl. I smiled. Gary's smile was all I saw. Lost in thought again, I heard the noise of G lighting another match and started in on another bowl of pot.

"It's the environment, Gary," I said. "It's just not the right light?" I rose and led Gary to the room's small closet. I closed the door, sat down, and directed him likewise. "It's darker in here. The room will hold the smoke better. I think that keeps you from getting high," I encouraged him another pipe. "Argghh! It's just not the right light," I spoke, jumping to my feet. Rapidly, I went to Erin's bathroom, and from a drawer, I located a pair of thin black woman's panties. Back inside the closet, I draped the panties over the exposed light bulb. "There that ought to do it." We smoked yet another pipe, and before it was empty, I'd no doubt that I was stoned.

"Hey, do you smell smoke?" Brute seriously asked. I quiescently gazed upward to admire black twisting smoke emanating from the light bulb. A finger of flame twisted off the panties.

"Oh yeah. The closet is on fire." Brute Force jumped up, tore down the underwear, stomped on them until there was a bloody paroxysm of smoke.

The digital clock insisted to me that we go out and leave this room, which was growing smaller by the day. Gary responded to my prodding by slumming further into his throw pillows. I considered the door a barrier. Going outside, I noticed to cough drop clouds against an otherwise azure blue sky. I looked and said hello to a young girl who was playing near the door. To the side were stairs, cement stairs. There must be at least a dozen of them. Taking one step backward then another, I ran forward and hurled myself out over the stairway. Clearing the entire stretch, I landed with surrounding smack, feet together *exactly* in pace. I turned around, asking myself if I or anyone could ever…

Soon, together, we walked in the crepuscular air amid musical notes of garbage cans being put out for the next day. I was not my usual self, for, despite the fact that Gary had grabbed me by the shirt and shoulder and pulled me back from a van that was making a surprise right turn, I put my nose into its side. The humor of this with Gary laughing cast me into the street where cars in both directions come to a decisive stop.

"Watch this!" I deadpanned. I put my arms out to form a T and put my tippy-toe into the street. Cars in both directions broke to a stop. Gary's eyes popped open just smaller than his open mouth hangs.

Apartment side, we had more pot. We were exceptionally hungry, but I refused to eat another pancake, and I demanded that we go out to eat.

"We just can't spend the money, Ned," Gary said sincerely. Then he distinctly said, "Tell you what. Promise me to stay put for half an hour." He returned after half an hour with his hands he held a Johnny cake, a large coffee cake. We made coffee instantly. Eating, although the food difference was exceptionally wanted after a week of pancakes and it was appreciated, I still wanted something. Reaching for the maple syrup, I put lots of it on my cake. Gary laughed. He laughed so hard he fell off his chair.

"What's so funny?"

"You are eating a giant pancake!"

CHAPTER 8

Sunset Strip

Sunset strip, the name for a sunset boulevard, was on a hill, overlooking a grand expanse of LA. We were here at the right time with the sun's setting rays—the atmosphere combined with the growing city lights flicker into a glorious, sparkling colorful sight. Being average teenagers, we toured, sat on fire hydrants, and walked the street, trying to be happy and entertained. Passing autos were crazy customized jobs with distinguished horns, chrome hats on automobile hoods, silver and reddish Statue of Liberty on a hood, monograms on the doors, rakes or slants two feet high pushing the nose of the cars almost into the ground.

Three girls in a passing car stopped and asked us if we want to go to a pot party. The girls won't go pick up a six of beer, so we refused. We watched people. I saw their short sleeves, Nehru shirts, turtlenecks, Hawaiian prints, sandals, beads, bent Cavendish pipes, thick high heels, miniskirts, walking sticks, canes shorts, and dungarees. The daily tabloid made me blanch, "Weather report—moderate smog." It began raining, and we made it home quickly to sleep.

I was awakened to see the first sunlight coming through green curtains.

"Boy!" Gary said pluckily. "I bet today is going to be fantastic. Today is our day." He wore a fine grin. Going for the door, Gary was met halfway by Erin just returning home.

"Hi, ahh, you've got to get out," she said. "It's my home, and I've been gone long enough. I'm sure I don't have to ask about that. And it is my place. Ahh, my boyfriend is coming over…in an hour, and I never bothered to explain you to him."

We guys each took a shower, and I trimmed down my goatee. I saw my sun-chapped skin was returning to normal, and I momentarily fretted about the crossing country in the deserts and the fact I'm not going to be looking any better any time soon. Lost in my reflection off the mirror, I snapped to hearing Erin and Gary arguing. We apparently had "won" a ride to the highway heading toward Mexico. Into her red sports convertible, I was hanging on for dear life. This time, she was trying to impress something on our minds.

About fifty miles would be a good lift, but we got to three hundred yards past a large sign reading Los Angeles and were let out. Immediately crossing both sides to the busy highway, we thumbed back into the city. The day was spent unsuccessfully, begging for food money.

"Shit," Gary said. "We spent a week here and haven't found any friends, gotten squeezed, or found work."

"Well," I voiced, "we going to just blow around, find a job, or what?"

"I'm for finding a girl to work for us," I spoke.

"Yeah, but your mother's three thousand miles away."

"Gary, which is it?"

"Hey!" he shouted at me. "Don't push me. How do I know?" We fellows had a way of finding out what would happen to us next.

"Hollywood?"

"Sure."

Gary asked for the price of a cup of Joe. He got the reply of "Eight-five cents, this ain't a dinner, darling." We did not have coffee. Garry got the ribs from me, a blow that must have hurt, but he said nothing. I exited the café with Gary, wanting to fight anyone but him. I suggested we split up.

Some women that I saw appreciate my flirting eyes, not that they were always going somewhere. I put my moves on one or two of the passing ladies while they moved away pronto. I started growing angry. I sat plunk on the plunk on the ground, arms folded across my knees, while I didn't give a damn about anything. A female flirted with me then scooted away. I felt like saying, "Screw you! You won't even split a glass of ginger ale." I was getting mixed up, and I didn't like this. A pair of women passed with their noses in the air, and I began to get mad at that. Then I saw that I looked ugly with my brow furrowed.

A woman exited a restaurant around noon. She stared at me a couple of times for a long time. "Take me home to work in your garden?" I said. The lady didn't need to check if she heard correctly, and she smiled, bursting out in laughter. I felt most embarrassed while my pantomime joke of patting myself on the back of trying got an appreciative knowing smile.

"Oh, honey," she said from long brunette hair and *not* wearing sunglasses. The woman stopped two female companions, who were exiting the restaurants, and repeated what I said word for word. The females had a prolonged laugh while I stood there lonely. The initial woman stopped laughing, turned to me, saying, "Oh, honey, it's just that what you said is so funny. It's not you. Ha, ha, ha. It's just that some of us have husbands, and telling them we have a Mexican gardener tells them more than meets the eyes."

There was no need for the lady to say more. I sorted of half smile also, but I had such a stomachache that is double knitted down. I attempted to leave as gracefully as possible, feeling angry at all women. A girl with braided hair held in front by a red kerchief about nineteen came over and asked me what the women were laughing at.

"Aw, I just asked them if they needed a gardener." The girl giggled. I knew she was giving me the come on, but I was so mad at women.

"The reason they're laughing is…" She broke out in laughter.

"I know…" I swore and muttered to the ground. She had pretty soft shoulders, and her blue eyes looking at my blue eyes was a mutual allurement. She was looking at me directly in the crotch. I simply knew she was interested in bedding me. And I sat there, knowing I was going to spitefully refuse her. It was strange.

GARY AND DUKE

"I've got some grass back at my place. It's not oregano either." She smiled. "Say, let's go there. I live alone, and we can work on it."

I felt attracted yet pulled back, saying really negatively, "Don't feel like it."

She was disappointed while she waited for the extra second.

Well, she said, "Goodbye," with a tone sharper than a razor.

I had blue balls on my very first step. I got mad at that!

CHAPTER 9

Jack and Jay

"Get the hell out of here! Who do you think you are? Who are you!" yelled an elderly bespeckled man with a gravel voice. I peered out of my sleeping bag and was blinded by the light of dawn, then I noticed the dew on the grass and an elderly man standing right over me. It was somewhere in LA. "I'm going to call the police," he said.

"Get the hell out of here!" I shouted back. The man went away. Then fearing the police would show up, I tossed away my switchblade. Suddenly, feeling defenseless, I retrieved the knife—the good old unknown. Through the back and then front lawns of several prosperous homes, we left footprints on the wet lawns. It was six ten in the morning when Gary wanted to do laundry. "You should see yourself," he said.

Down to trousers, when a policeman walked into the Laundromat and said, "Nooo!" Gary or I was about to drop our trousers. Almost exposed, when we left Erin, we had arranged to purchase a lid of pot for fourteen dollars. This was put in a slit in my jacket pocket into the lining.

I'd decided to call a never-met relative, my mother's cousin, Jack Conevy, who lived in LA. I figured that at least we could get a square meal in.

"Meet you in twenty minutes after I shave and have my coffee," Jack said. And he appeared at the corner with a new green Mustang, top-up. He exited his car, came over to the telephone booth, and introduced himself while casually mentioning he is vice president of the 'X' bank of California. Then we drove three miles to Jack's apartment, where we boys had our third cup of coffee in about as many weeks. He shaved, telling us, "I'll take a taxi home. You can have the car. Be sure to eat a really good meal, although I think there's only bacon, eggs, and toast."

I drove with Jack to his business in downtown Burbank, almost getting lost on the return trip. There was eschewed bacon for breakfast. I was inclined to go for a walk when the car was remembered, and we headed north along the coast.

"Why the hell north?" Gary asked.

"Because I'm driving and I wanted to see some of this highway one that people have been talking about."

We picked up a hitchhiker, who right off the bat bragged, saying he was thumbed for two hundred miles. All the while, skies were filling up heavy with what appeared to soon be rain. We drove to Malibu. Surfers boarded twenty-foot waves just offshore but hundred yards from where we drove. The waves were daunting, and we were quite impressed with the surfer's bravado. And we both agreed we wouldn't dare bodysurf in the water such as this. As we dropped off the hitchhiker, we went half a mile further, switched driving position, then laughed at the look on the guy when we returned going the other way.

Returning to the apartment, we fought battles with the chairs. Gary opted for the floor. While we read the *Times*, I somehow noticed an ad in the help wanted section, "Wanted a professional male wants company to drive to South Caroline. All expenses paid." With the phone against my ear, I asked the operator what day it was.

"Monday."

"Hmmmm?" Gary looked up from the funnies, where he was enjoying a laugh.

"Whadda ya think of an all-expense paid for a trip to the East Coast?"

Gary exhaled aloud then wistfully said, "I wish it could be." The ad was torn from the paper, and the number was called. The party was leaving soon today, and two drivers would be acceptable.

I left a note on Jack's table, saying, "Thank you."

We young men met the party within two hours. The driver introduced himself as "Jay." He got a crimson scar encompassing his neck. In the passenger seat was a woman yacking away even while we were being introduced. We followed steps to the rear of the car, where I actually stated, "Let's save Jay from her."

"This baby has got a full eight," Gary said. The car was a two-year-old tan cutlass. However, the car was fully loaded with these two people and their belongings. The female was really something else, a real beauty. Nasally complaining, she, obvious to me, didn't enjoy Jay's company. She chastised both he and us at every turn, interrupting every sentence with "Shut up, I'm talking!" she was a real pain in the neck. This fellow, Jay, was much harder to size up. He responded to the woman's every sentence with needless words, feeble defense. In addition to a real neck scar, this Jay had a face quite pockmarked. Still, I felt sorry for him and exchanged glances with Gary agreeing. The woman just won't stop loudly complaining.

"Stop this nuthouse! I want to get out!" I madly expressed to the car.

I was left with two suitcases, a record player, and a package on a small hill. "Keep the faith, kid," Gary said while leaving. Gary headed for the sunset while I was left alone with this and that while people drive by me and give the peace sign. I was a guy who looked like he was on his ass, down on my luck, sitting on a hill waiting for time itself to explain.

A faint whisper of wind touched my forehead, and I looked up to see an elderly lady with her cane just turning from me, having said, "Heavenly God bless."

After four hours, Gary and Jay returned alone inside their cutlass. The rear seat was covered in plastic clothes wrappers. This unnerved me, and I clenched my fists tightly.

"Ned, c'mon. I'll explain later. Let's just get started," Gary said evenly.

"This is going to be a hell of a conduct," I said, moving over some of the plastic wraps. Jay put the car into gear, his hands shaking, fingers running through his hair, wheelhouse turning the front of the car this way and that. He said how sorry he was for how things have gotten started.

"Look, I've got credit cards, and I'll be buying breakfast and dinner every night, plus I'll pay for lodging," Jay said, stammering. Then he continued, "All you have to do is drive every now and then." I said nothing, and Jay grew more frantic. Finally, without much pop, he asserted, "If you don't like it, then you can get out!"

I half said I was sorry then did say, "I've got a good reason for being angry. You were gone for a long time!"

"Okay, I'll start over," Jay said. "I'm Jay, and I've come here from the Carolinas. I'm a schoolteacher, and I was here on my vacation." Gary rolled his eyes skyward. Jay continued, "And the people I was with took me for everything I have…drink, motel bill, including them, food, everything." He paid emphasis on his words, but he was a nasal drone.

"Why?" I said. "Didn't you just drill her with some husk?"

"Her?"

"Hmmm, good point," I expressed. I was just not "into" it. I expressed that all I wanted to do was cross-country. "None of business, understand?"

"She," Jay said, "was such an asshole." He pronounced, "Her husband too. Assholes."

"All right, I've had it!" Gary screamed from the passenger seat. "Had it! I don't want to hear it a fifth time! Ned, I'll explain it all later."

"Okay," I said.

"Thanks, Savage."

"Savage?" Jay asked nervously.

"His nickname. Forget it," Brute sizzled.

"Do you have identification?" Jay asked. Gary and I broke out in laughter.

CHAPTER 10

Headed East

As we drove away from Los Angeles and her suburbs, it was growing dark. Casas or adobe dwelling became rare, and there were conspicuously fewer and fewer lights, which almost made me want to count them. We entered Arizona through a border crossing, where we did not declare any plants or weeds. Gary switched driving positions with Jay just east of the coastal range and Sierra Neva at a sea level of nine and a half thousand feet.

Oh no, I thought as I saw Jay's left arm touch the console, then he rested his arm on the front seat divider then slowly ever slowly reached for my knee.

Much as I wanted to see the Mohave Desert, from the car, only flat grayish sand without any fauna was seen. About midnight, we arrived at an inexpensive but sole motel that afforded us a double bed and a single. I was starting to think how Jay wasn't so bad; however, when we get to our room, Jay hopped vigorously onto the double bed.

"Oh, boy," I said to Gary. "Here we go."

After each using the toilet, Gary and I unfurled our sleeping bags onto the floor. I was glad the floor had a rug. "Hey, you fool, what are you doing?" Jay stated with a vigorous high pitched voice. He patted the bed like some elbow strained football player. "I paid for these beds, now c'mon," he averred.

I came up with "We're so used to camping that we couldn't sleep in a bed." Gary nodded in agreement.

"I said c'mon," antagonized Jay said.

I was thinking about how I didn't know where I was. I felt the bristles to my beard stood erect. I gave Jay a singular look, and the matter was settled. Jay spit a few gruff words made funny how he said them. He rose from the bed. Is he leaving? Jay went to his luggage, extracted three issues of *Playboy* from his bad, and entered the toilet, slamming the door shut. Some three hours later, Jay came out of the toilet. He looked at me and dropped the magazine onto his bed, the motion of which caused his towel to slip from his midriff, exposing him as totally naked. He smiled coyly at me. I groaned. Then and only then I went to sleep.

At daylight, Jay said, "Gee, I hope we have enough money to make it cross-country. It's a long way. Sure is." This made it perhaps the fourth time since departing that Jay speaks similar words. Jay ordered breakfast, and I made sure to order bacon. Accepting a coin toss, I must sit in the rear seat again.

We drove a pair of hundred miles listening to Jay rant on and on, where even the sounds from the car's engine sound better than his nasal going on and on. He would not even turn on the radio.

"No. It's my car, and I feel like talking," he spoke.

At a restaurant, Jay ignored the waitress and sauntered around. Then he went to the take-out counter. I laughed at my stomach noises.

"Twice I've wanted to belt him," Gary said.

"Gee, I hope I have enough money to make it cross-country," Jay said. I stared hard at the jackalope postcards.

Naturally, I wanted to sit in the front seat; however, Jay said how he wanted his "amigo" Gary to drive. "No. It's my car, and I want to sit in front. I don't trust you. If you don't like it, walk." I felt like a bottled scorpion.

"What do I care," I said, tearing down the plastic clothing wraps from both of the car's rear seat windows. After a pause, Jay swallowed several pills.

"It's all right. Just stomach pills that I got from back home…for my stomach." Gary looked at me with concern at what Jay had said. I lay down sideways on the back seat. In time, Jay had again swung his arm around, and it came to rest on my knee. I grew mad. The further down the leg he went, the more I was ready to burst without wonder. I looked around and saw underneath Jay's front seat a penknife. I opened it, about three inches long. Jay's hand got to a certain point on my leg. And I jabbed him with the penknife. Jay didn't flinch.

I touched the knife, finding it reasonably sharp. This time, I forcibly pushed the knife into his arm. Jay removed his arm *slowly*! I regarded that he had taken some opiates, and about anyone would believe me if I tell them this true story.

Soon, Jay said he must go visit a doctor. He had digestive trouble. We argued and drove to an MD. Once visiting the doctor then taking a trip to the pharmacy, Jay ingested about seven pills before we had time to stop him or utter a protest. Gary and I had been duped. Jay became giggly pooh. Then he fell asleep. Gary reached across the way and put his hand underneath Jay's nostrils, finding him breathing.

"Yeah, yeah, he's sleeping."

I was wondering, *Phew, ya' know if anything happens to him, it's our ass.*

"What a creep he is!"

Gary said, "At least you can fall asleep. I have to sit up front with him."

"I think he likes you."

"Oh, drop dead, what do we do, just let him sleep?"

"I guess so," I stated. "Just check on him from time to time."

"Man," Gary expressed loudly, "we've got to get on top of this."

"You won't believe this. But a while back, I stabbed him a knife, and he didn't flinch."

"Nah."

"He didn't. I did."

"Oh, come on now, your knife?"

"No. Another one." I showed the knife to him. "And that was before he took the other dose. Well, then," I pronounce, "We sure as hell better keep him awake." I shoved Jay against his door.

"Hey," Jay said from far away.

"Hey yourself. You stay awake. You hear me?" I said to Jay, who smiled affectionately.

"Why don't you save some of those pills for later tonight with Gary?"

"You want me to?" Jay enlarged his eyes and asked Gary.

The moment that Jay fell asleep, Gary said to me, "Let him sleep a few minutes. I don't want him to see what I'm doing. I wouldn't be doing this if traffic wasn't so heavy and slow." Gary drove, crossing our lane and going into the approaching lane. "No passing" signs, we flew past in the auto. Ahead in view in the westbound lane was an Arizona state trooper car. This other car sped up noticeably toward us. When we moved into our traveling lane, we had missed the cruiser by a split second—Gary doing ninety and the trooper doing about a hundred.

"Crazy son of a bitch," Gary swore. Together, we looked back for the trooper who had pulled a U-turn and accelerated so that he had almost three or four other head-on collisions.

"That guy is totally crazy, Ned." Five miles from where we first approached the state trooper, he had pulled us over to the dirt edge to the road—lights flashing and a damn horn on and on.

"A ticket doesn't matter," I said. "We can just skip state. Just don't get him angry."

"I won't pummel him," Gary said straight, fingering the steering wheel, muttering.

This pair of mirror sunglasses, the trooper, six feet three in shiny black boots, asked for Gary's identification and the car's registration. I went toward the front of the car for the papers—only Jay was unconscious against his door. I rapped on the window, but it did no good. When I opened the door, Jay's body fell down hard into the dirt. The trooper drew his forty-five and declared, "Up against the car and spread 'em. You are under arrest for manslaughter."

I was frisked. Then the trooper treated Gary contemptuously, yanking his arm, making aspersion about his looks and brains. They were busy, and I went over to Jay's body, which was dangling half in and half out of the cutlass. Jay had been sound asleep through the bustle. I slapped him on the face a few times. He responded with a low, slow moan.

"Get away from him!" the trooper ordered. I slapped Jay again. He responded with a low moan and some indistinguishable remark about his face. "He took some pills!" wide wild-eyed Gary said. "Seven of them," he said. "We tried to stop him, but it was too late because he swallowed them! What? Are we responsible for a grown man?" I cringed at these words.

The plain-faced uniformed statie said, "Where is the bottle," behind mirror sunglasses. The bottle, very strange, bore no marking except where it was obtained.

"You're gonna follow me, see, behind my car," he said, sheathing his weapon. "And when we got to the doctors you're lying about, we're going to check things out. Drive more than one car length behind me, and I'll immediately take you to prison." Fifty miles back, never altering from the troopers fifty-five miles per hour speed, Gary and I knew damn well the trooper believe Jay was on sleeping pills, and all we had to do was keep him awake.

We arrived at the doctor. In the parking lot, the trooper told us, "I'm going in to see the doctor. Leave here, and I'll fry your asses black. It's sixty miles to the border, so don't even think of disappearing. Stay put you and be certain to keep him awake."

"Jez, what a dipshit," I remarked. We couldn't wait more than ten minutes, then we headed for the doctor's office just as the state and doctor were coming out.

"Just a mild sedative, boys, just half the dose of what he asked for," said the MD. "Just keep him awake a few hours, and he'll be okay. If he sleeps a five-minute stretch, it's okay. Just be sure he's breathing." The physician gave a friendly smile, as I noticed the trooper's mirror glasses do not reflect the daylight growing to night. The trooper put his boot on the car's bumper, and he wrote out a ticket for Gary "splitting traffic" him (going a hundred mph).

"I'll see you two late," said the trooper, who inched down his sunglasses to reveal two uninspiring brown cold eyes. Gary bit his tongue and looked at me. I said nilch.

Finally, it was my turn to drive. I slapped Jay's face on this hundred-degree day. Jay fell asleep again. I adjusted the driver's seat, then I shoved Jay against the passenger's door.

"Oooh, what a dream," Jay said. "It was so vivid and real." What's going on was Jay thought we were leaving the doctor's parking lot for the first time. He'd lost three or four hours in an instant. Gary and I did not give out any word.

I suggested coffee, stopped the car, and returned with two coffees and a cup of water. "What's the water for?" Gary asked.

"Jay," said and tossed the cupful flush onto Jay's face. Jay was awakened, babbling, smiling, refreshed. I sang a soft lullaby to Jay then blasted the radio, saying, "Jay, your pants are on fire! Jay, wake up!" Jay didn't mind and smiled. Jay had his own spot, as he pretended to accidentally plop his head onto my lap while I was driving. Then I bounced him off the passenger's door. What about his glasses? Nope, I missed him that time.

Towns were thirty to forty miles apart in this part of Arizona, scared desert—dust, even dust yellowish, beige, white on occasion, and a few cacti or shrivel tree. Where a river had once existed, flowed, breathed life into the earth, there were clay and hardened sand deposits giving earthen red color sand deposits—such a cousin from Georgia's red earth. There was exposed earth in levels—a magenta hue, maroon, ochre, rust, dark pink. Oxblood showed and this about the seventh shade of red here. I was able to spot the scene and waited for nature to do her show. I showed her ribs by the stripping exotic hypnotizing setting sun's light.

The notion of the desert here as painted was not nonsense as the sun began a two-hour-long descent into the desert floor, yet as the desert drew from prior changes in the river level and its contents became accentuated shadows of no diminutive concept—eons of presence being exposed—came to glittering, hueing, terrestrial magnificence. That a canvas might not capture the picture was a diminutive concept. Penetrating rays drew color in and out of the rock. Daydarkened curved canyon caved on the walls, perhaps animal lairs, deep with the colors of a former day's river treasures, become for this one time of day revealing where notions of the phoenix prevail.

An upper torso was chilled. My arm stayed put on the door ledge as it gathered final sunrays. Radio was long off. The other fellows were snoring. Vigorously, I shook Jay or tweaked his nose to make sure he was breathing. I lit a cigar and all, but my driving was ignored. The `sun set as if an ember in a pipe bowl, a fierce seething ember settling, consuming all imagery. Alien colors came out from the rocks, and the sun seemed to turn blood red. Rocks cast silver, permanent neutrals, sands, reds, all precariously grabbing my tired thoughts and leaving me unable to think. After several times of a dark void, the evening stars brought out the night light. Being hidden in flatlands, the virulent sun rays had driven me wild. And I dreamed of being driven inland by rustler escaping and then avoiding being killed by a wild bison on the untractable land. Then, I had a variation of my incubus, and the rattlesnake poison had not set in time, and I was able to knock the lone guard out with a rock, got to the sheriff's, and saved Gary before the venom he had also been tortured with kills him. The evening was black with only a soft shadowy gray to see by. I saw a majestic tepee at the edge of the plateau.

As we entered the next town, it was well in the evening. I'd decided we'd get a room with a bath, and that was that.

"Let me drive," Jay spoke. "It's my turn, and I want to keep going. I've been sleeping. You two guys can sleep in the car. Come on, I want to go another hundred miles at least," Jay ranted at me. He was ignored.

Booking to the room with double beds and a tub, Jay said he was going out for a cup of coffee. By the time his ludicrous statement registered, it was too late. I knew his car had gone somewhere into the night. While drawing a bath, I said, "Go get some comfort. There must be a bar around here. Get it over the counter."

"Hey, I got the booze," Gary said with an accomplished half grin.

"Well, hels bells give me some." Gary greeted my outstretched hand with a large plastic cup filled with ice cubes and southern comfort. But Gary was doing what's unusual for him sidestepping. "Ahh, well, ahh. I went into the office for some ice cubes, y' know." His voice echoed off the bathroom's walls. "And, ah, I found my gear. But yours wasn't there." I stared at the two bars of soap just above the waterline in the tub and crushed them into a mass.

CHAPTER 11

Rocking in the Sixties

"Hit the road, time a new day, and we've got to get going before the sun does beat us to pieces, that is," I said. I directed Gary, who was just getting out of bed and rubbing the sleep out of his eyes. We tossed a few jokes and punched at one another. As we hit the road, the sun was not high. And it might be nine thirty. After an hour of what promises to be another hot day, I looked frantically for a hand-sized rock. Jay drove by.

I screamed madly into the air and waved my arms wildly. A moment later, a family, complete with a son, daughter, and a puppy, drove past and waved hello. I waved back to them. After an hour, we drove a five-mile ride. Let off at an old Texaco gas station, we purchased two water bags. We wanted for any discarded cardboard to beat the heat but were refused. A thermometer on the side of the building pointed east and registered 118 at eleven o'clock. Crossing the road, we had a ride in fifteen minutes.

Gary right away poked fun at the driver who, with his passenger hail from the state of New York. Their accents were pronounced clearly and sharply. Gary didn't stop with his mocking, and after three miles, the car screeched to a hit, and we were told to get out! Their car was rolling away with the license plates conspicuous. Kicking at the dust, Gary remembered that our water bags were in the trunk of the car. I howled. Punches were thrown at each other until the heat made a quick impression. Seven hours later, toward the end of the day, there was no spittle left in our mouths to put on overly parched lips.

Just a good size rock was finally seen out in the desert after these hours of thumbing. I paced forty feet into the searing desert, which was heat vibrating and white how away from the road. I carried the rock torso cradle back toward the road (noticing the rock holding me to the ground). I spotted a pickup truck half a mile away, *and I howled aloud!* Gary looked at me and thought I've slipped. I laughed and laughed some more when I held the rock close, and the pickup truck stopped for us. When I got into the cab, I waited for the driver to speak. The white-haired middle-aged man smiled cordially and said, "I saw you there in the desert getting the rock, and I didn't want you to get comfortable."

We could go with the driver to Chicago if we wanted to. The cab was a customized cabin, and a compass rested on the smart tan dashboard. I lost the air-conditioning argument (it was left on).

"What's the compass for?" Gary questioned.

"Hell, gentlemen, I never know when I'm going to take this baby off the road or highway. Why, at any moment out west here, you just might spot a rabbit or two and spin off the road and try to pop them." He pulled back a yellow curtain to show the cabins sleeping compartment and mounted on finely carved wood above that are four riffles. After some forty miles, the cabin driver said, "Tell me which way you're going, guys. The road separates in a few miles. I'm going north, or do you want to go straight east?"

"Route sixty-six goes through Oklahoma, doesn't it?" I asked.

"Yes."

This pleased me as a favorite uncle, my uncle Ed Perkins, was hailed from Oklahoma. For the ride, we chatted about guns, the weather, and a chat while sitting on something was super.

We trod a mile without cars passing. Then at a rare intersection, there was a solitary bakery where we stood in front of the plate glass and imagined smells emanating from behind the glass. On the other road of this intersection, we stood in a place listening to the world when a large blue station wagon with a man driving, a lady passenger, and kids in the rear. Zoot went to the electric window that let air condition air escape. I must step back three steps to avoid the contrasting temperatures. The driver bent over from the steering wheel to bypass his passenger wife to say, "Are you looking for." I considered that the word he used was "grub," but I couldn't be sure.

"Yes," I said to the woman. "We're looking for the corner of Elm and Main." The woman scowled, the man frowned, and Zoot the car went away. I laughed at what I had said. Gary decided not to bop me for it, and we howled at it all.

Right on time (smile), two girls our age drove to the stop sign at the intersection. They smiled at us. They smiled, basking in the sun.

I made a crude remark about western meats, and the duo sped away, leaving us completely alone.

After half an hour, an old dated gray two-door sedan pulled over, and the driver waved for us to come on.

"Hi, how you today?" Gary said energetically, pumping the man's hand. The stranger returned the grip, and he changed the standard auto's gears using his left arm and hand. Well, old Brue Force and me. El Savage, we started up a barn burner of a conversation in that small way, paying back for the lift. The driver clucked and nodded. A couple of times, the driver pointed at the speedometer and pointed at the dial at sixty-five. I figured we'd be going sixty-five once we got up steam. We, two hitchhikers, enjoyed ourselves immensely. The sound bounced off the car's roof and windows. In a couple of minutes, the driver reached behind to the rear seat's floor, where right by my feet, he extracted something from a paper bag. He handled it to me. It was a small hand-carved figurine of an elephant with a memo, "I am a deaf-mute. Thank you for the contribution." Yes, twenty-odd minutes, Gary and I had been carrying on a conversation with a deaf and mute individual.

The old Ford drove at forty-five, peaked at fifty. The sought for sixty-five was just that sought for. It was a long way to Oklahoma City because the car broke down. The bald driver got out, opened the hood quite efficiently, then looking distressed, he put his hands on his hips. The man's face would be funny under any other circumstances. Gary was a real Johnny on the spotting mechanic. After singing one of his arms, he got the engine running. But it stalled out. Gary went to the front of the car, put his hands on his hip, and jumped up and down. The drive looked questioningly at me, and I thought Gary needed a larger Stetson. Gary came forward, pulled the driver away from the ignition's front seat. Gary again jumped up and down. Then the driver jumped up and down. Bemused, I appreciated one man communicate to another that they're going to jump-start a car. After one try, Gary went to the front seat and released the emergency brake. Then, we jump-started the car.

In two miles, there was a gas station and diner. Gary borrowed tools from the mechanic there, and Gary got busy for two hours. As he finished, the car's motor sounded just fine. Couldn't locate the owner, I walked into the dining to hear, "Who does he think he is, ordering then refusing to pay, showing me his empty wallet! I knew from the looks of him. That's why I asked if he has any money." It was told by a very angry waitress to a counter full of a patron.

"How can he order? He doesn't even speak," I said to an onlooking room that burst into laughter.

He pointed at the picture. "That's how, smarty pants!" the waitress snapped. She was not about to pay for the large hamburger. Gary paid for the man's order. Only he took three-quarters of the remaining burger for himself. Our driver remained in his booth alone, shaking his head. Either the price or the hamburger disagreed with him.

"Give you half of what's left for eighty-five cents."

"You're nuts, Gary. Still, you helping him out is pretty nice." I teased Gary by saying, "Don't you figure we owe him something? I mean, he is selling trinkets to just get by."

"Who cares!" Gary mumbled through a full mouth. "I have just spent three hours fixing his car...got a scorch arm for it. I have never seen him before and don't care if I ever see him again. And you want me to feed him? Well, forget it!"

We needed to carry the poor chap out by his elbows that got a number of laughs. Back in the car, the man said, "Hmmmmmm, okkkkay."

I said, "Oklahoma City?" And the man nodded. He again pointed at the speedometer at sixty-five. We traveled five miles with the man, and the man drove to a side to a dirt drive. He exited the car, opened a gate, and waved bye at us.

"Unnnnh," he said to Gary, straight-faced. Gary sought help, and I deadpanned to him.

"Hunnnnnh." The man pointed at his odometer again. After two minutes, I saw it clearly as a picture. The guy was trying to say, "Home." Oklahoma City just happened to be sixty-five miles from where he picked us up. (It was six and a half miles we had gone with him.) Okohome-ahh! It was night, and we stood there by the swinging gate, looking at the stars feeling alone. There was no smell of sirocco here. Here the celestial sky was lovely.

"Lordy!" Gary said to me. "Sure hope that burger does not sting like a pellet. Cost me more than money." We wanted to pound the crap out of the other. Footing it maybe three-quarters of a mile, we tried thumbing. The sky was clear. The moon was full—no howls of any critter. I used the horse blanket, pulling it up to my neck. I lit both cigars I had bought at the diner. I watched the cigar ember. Later, the moon clouded over, and about four or five o'clock, the sun notched up from the desert ever so slowly.

Forgetting my prayers, we spent two hours alone by the edge to Route 66, single lane in each direction. Rarely, I mean very a truck or ace drove by. Then we ignored a host of semis when it was exactly the tenth car that stops.

"If you all fellows want to pay for gas, you can have a ride," said a goateed and leather-jacketed maybe thirty-year-old man, sanding outside his autos so early in the morning. He wore a jet-black T-shirt. He was operating a 1956 Thunderbird.

"That a fifty-six or seven."

"Fifty-six. Do you agree?"

"Yes."

"Okay." The man took off his jacket to reveal his slender arms and look of experience. "Okay, out," he said until the instant then noticed an old bearded man in the passenger's seat. This passenger looked like an old prospector without his ass. His hands' nitty-gritty creased and looked massively aged, fingernails curling and half-dead yellow. His overalls were dirt-caked and covered with dust. He slowly shifted his feet in the desert sand.

"That guy's worse off than us," I said to Gary.

"Yeah. I agree."

"C'mon," the bearded driver of the car spoke.

"This is bothersome, Ned. I mean, he's in for a battle of survival, too, because inside an hour is going to be boiling. Whatcha say we refuse this ride?"

"Look at me." I retorted. "He doesn't even know where he's going. Let's give him two dollars or something. Let's just apologize."

"Okay," dropped the car's owner. "Let's get going." Right at fifty miles later, we arrived at a gas station. I had wanted to cry because sitting on the console was so painful to my ass. The car's suspension gave at every bump. We handed over five dollars for gasoline then burned down a few peanut butter crackers with a Coke.

The five-foot-ten bearded driver spoke, "It's your turn to drive, Gary." And then he comfortably got into the car. I couldn't stand it.

Then the owner said, "That console must be a real killer. You drive, Ned." Then the song "Zorba the Greek" came on the radio. This hombre, he looked quite tired, and on the stinging console, he soon fell asleep.

And then the song "Zorba the Greek" came on the radio. I *thumped* on the gas pedal synchronously to the strongly downbeat the music. We wiped tears from our eyes. We laughed so hard. The music ceased. The owner woke up, saying, "Slow down...I'd just a soon get where we're going."

At twilight, we stood in place, thumbing. A car tooted its horn. I, with a nasty look, turned to face it, but it was that all-American family keeping pace, waving. It got so dark that we stood sideways with shirts stuffed underneath the top of our T-shirts to appear in the night to appear as females.

Noon, we had given up thumbing, having eyeballed a sign reading, "Howard Johnson restaurant five miles." We reflected, walked, rested, and walked. A car with a rumble seat stopped, and we went seven miles until the next town. Out west, the laws generally read, "No hitchhiking within city limits." We walked past a stop sign and expected to walk through the town. Only a garbage truck driven by a teenager wearing mirror sunglasses stopped and said, "Going in a mile or so, just thought I'd help you out." This co-teenager who drove the truck sparkled with enthusiasm, picking up our spirits. Gary and I were punching at one another after being let out of the truck when Gary grabbed me by the shoulder and won't let go while trying to turn me.

"Hey, watch it," I said, moving away from what I expected to be a loose punch, yet I saw over Gary's shoulder a parking lot that he was pointing to. I looked but didn't see anything in particular, only parked cars.

"Look. Look!" Gary said. Maybe forty yards away was Jay's automobile—his model, his color, his tires, and his plates. Never minding our scallywag appearance, we walked directly to the motel's lobby and asked for Jay's room number, saying he was expecting us.

Knock, knock.

"Who's there?"

"Sir, this is room service. Open the door. I have a message from the front desk."

A nasal whinny said, "Nobody could unless she...but how could she."

Jay spoke as he opened the door, "I don't know who it can be." When Jay saw us, his jaw drops, and the lone towel that is wrapped around his waist dropped, leaving him naked, and he used that for a pass.

I groaned and said aloud, "I won't touch him. I won't touch him," while arms tightly folded across my chest. I glared.

"Where are Ned's things?" Gary demanded strongly.

"Whose?" Jay said in a very stupid way.

"Ned's! Where in hell is his gear?"

"They're in my trunk. I swear I didn't know I had them until last night."

"I don't care what. I don't give a damn what you say!" Gary shouted. "My gear was next to his when you left mine *outside* the motel office. Give me the keys."

"Sure. Sure. I'll get the keys and his stuff. Just let me get dressed first." We made the error of stepping backward outside of Jay's room, letting him lock the door and chain it.

"What are you going to do?" Brue Force asked me. "Break down the door?"

"Nah, I'll go see the manager," I replied.

"Wait a few seconds, Savage. If you saw yourself coming down the street, you'd scream."

I did understand that if I looked at myself in the mirror, I'd act differently. I actively decided not to look at myself. The hotel manager was quite affable. He invited us into his office, even offering me his own desk chair. Making a point of looking at the man right in the eyes, I began my story. The manager interrupted me to say, "Wait a minute. I'll get him." He reached for the phone's receiver.

What ensues was the manager got Jay on the phone and handed me the receiver, while Jay remained to think he's continuing to talk to the manager. I shot Gary a look, and he kept the motel's man occupied. I worked on the conversation.

"Mister, I think these boys have a pretty good case, and I've half a mind to bring the police in on this. It is robbery, y' understand. And I just might abet them saying you're saying through this motel *naked as a bartop dancer*!" Jay was flabbergasted. He must think cameras were in his room. He was not outraged. "Now wait a minute…I'll cooperate. Do you think some cash might help?" Jay stuttered.

I looked at the office and began to smile.

Gary, Yankee-doodle Gary, had it figured out.

The manager offered us use of the motel's station wagon to get us to the city limits when we are done.

Knock, knock.

"Room service," I said with a disguised voice. "Opin de door and put de keys on de towels, misstah?" The door locked was opened, but the chain remained. We stepped back from the door. The sound of the chain was heard, and then the door was kicked open, missing Jay's nose sweeping in.

The unlocked trunk exposed my gear, Aussie hat, and about eight issues of *Playboy*, those which Jay put into my outstretching but bewildered arms. I just stared and blinked. Gary took the magazines and bounced them off Jay's skull.

The air-conditioned station wagon was simply heaven as it drove us a few miles to get out of the town to a few hundred feet past a great spoon's aromas.

I was thirsty for more of the water that we got at the motel's bubbler, between Jay taking off with my gear and this location—here five hundred miles.

CHAPTER 12

Hitchhiking Some

Even at the break of day, I had a dull ache in my head.
"You'd better not do that," said Gary.
"Vy not?"
"Vy not? Because when you bend zee away from ze coming trucks, you make for ze nicesa target for ze backfires."
We thumbed for two hours then rested on a small hill beside a truck stop and diner. We couldn't stand the road. For an hour, we thumbed while sitting down on the comfortable green grass. Gary rose, saying, "Going over to the pumps, see what I can scrape up."
"Good thing it's not Thanksgiving," I rejoined. Ho, my buddy did it! He talked for five minutes with an older, bespectacled man. Gary gestured with his hands, used a smile I could see at fifty paces, and he got us a lift to Oklahoma City, a fine stretch of road.
Along this stretch of road, I was expecting more desert and tepees but was set to change my mind by seeing some spacious and fertile hills in Oklahoma by route sixty-six.
This driver was uptight as he fidgeted his wrinkled hands on his Chrysler steering wheel—avoiding conversation with Gary, who was in the front seat. We were inside a luxury Chrysler that's air-conditioned and telephoned (1968!) and had thick wall-to-wall carpeting. I was extra conversational and amiable toward the man who told us he was a pediatrician and a cattle farmer. I found this humorous, and

I thought of my girl, who summered by the ocean edge near Cape Cod in Massachusetts.

"Yep," the doctor said. "A canal goes to link up Tulsa to the Arkansas river then the Mississippi and the ocean, of course."

"Of course."

This was when car phones were very rare. The phone rang, and a pregnant woman was given advice by the doctor.

The chat subsided, and I was given to thinking about all the various rides Gary and I had shared on this trip.

"Hey, a limousine!" I shouted into the insides of this opulent quiet car.

"What are you talking about?"

"A limo, about the only thing we haven't been given a ride from is a limousine."

"Nope. Had one, remember, from the beach, not the guy from San Francisco who passed us several times."

"You remember him too?"

"Sure 'nuff," I asked sixty or seventyish doctor about oil wells.

Honestly, I don't know how interesting this is, but I'll just continue on with the story.

About nine or ten o'clock, we skinny-dipped at the motel's pool. Opposite us, at the shallow end were two couples in their early twenties in their bathing suite, who all smiled at us. Gary and I wrestled one another off the diving board, pulled on our bottles, and were loud enough for the manager to ignore our bareness and tell us to be quiet.

A few snorted back in the room, then a quick scuffled on the bed. Gary was rough, and I admitted defeat. I tried to go to sleep while Gary watched Johnny Carson (JC Superstar) and then a Japanese monster film. "Ha, ha, ha. Look at that, will ya?"

Awakening and letting Gary nodded off again, I put on well-worn dungarees and barefoot it outside where I startled the cleaning woman who sucked in her breath at my dishabille, unzipped trousers, no shirt. *A bit of contrast to last night*, I think. *Well, the worst is over. What could be so bad as the desert and sun?* My stomach grabbed me in all-consuming pain—for about a minute. It had been about a meal every two days now, and it was pointless to make any more jokes about acids that gnarl at my intestines. Putting my hand inside my right Levi pocket, I discovered a hole, a two-dollar missing. Still, glad to have divided the money into separate pockets, shoes, and socks. Returning from pacing about, I awakened Gary. I gathered from a morning newspaper that was Saturday.

Gary snapped to punctiliously, "Got to get going. Let's rush thing." Gary feigned a body check into me and left the room with his shirttail and I following.

"Well, after Texas, the sun won't be as hot," Gary said.

"Let's hope so," I replied evenly.

"You were the first one up. You should have thought of drinking water!" Gary yelled at me in pointless prick point aggravated by then start sun still climbing in the sky, the pulsating sun.

Blisters on the face lacked moisture—just have to thumb. We climbed over a guardrail and walked just to get out of the sun for a moment to a state highway department building. The plumbing in the edifice was not working, and there was no machine for tonic. Ready for the road. Not so ready.

"The campers, Ned! The campers!" Shoulder grabbed. I had goosed the information that blisters on our face lack moisture. Walking to a toll booth area, we just thumbed. To get out from the sun, we walked to a state highway edifice.

"The campers, Ned! The campers!" Gary pulled at my sleeve, goosed me, shoulder grabbed. I got the information that the campers parked outside the building might have cisterns with water in them. I was face-to-face with a man who was exiting his camper rear door.

"Mister, have you got any water to spare?" The man refused. My head held high. I paced toward the toll booth.

"Why, George? Why, George? How can you?" a woman from the same camper said. "Yoo-hoo-boys! We can help you. Oh, George, we have the sodas in the refrigerator." The woman held a can outstretch in her hand. We were each given a can of tonic, and my facial muscles relaxed.

After an hour, a coin flip decided who would get a ride first and get out of the sun. Gary was good and quiet about winning. He walked half a mile away, and I saw him get a ride. I digested a small pebble for geophagism for the alimentary tract.

At night, a lengthily blue station wagon pulling an almost equally long camper, it was spectral. Its taillights were illuminating the darkness. After some nine hours, I was awash with expectation. Who will be stopping? I, in my purple with a gold stripe football shirt, got into a station wagon that provided comfort to a man, a woman in the passenger seat, a young teenager in the back with his eight-year-old sister.

The man greeted me from behind the steering wheel, "What the hell are you doing out on the road so late at night? I almost hit you. I was breaking only to avoid hitting you. They got me to stop." Mother steered the conversation to better things. Harvey, the teen, got me to wolf down some cheese crackers. "Thanks, I'll have a few."

The youngest played with the fringes on my pants, carefully eyeing my boots. I answered her questions, "Of course, I'm not alone. I'm just not with him. Well, my wife is at home. That's what happens to your face when you get as old as me." The next to last question seemed to make Dad friendlier by the way he looked at his wife in the front seat—a quiet look that said a lot.

I attempted to sleep by the pavement. In the morning, after an hour, a white Plymouth convertible with its top-down crushed gravel to an atop. The clock read six fifty-five. I appreciated the maneuver the car produced to stop for me, but when it got going at ninety and held it, I didn't know what to think. As we started to cruise, I saw Gary sleeping atop a guardrail. "Ha, ha, ha. Look at that dumb SOB," said the driver, who held the wheel agitatedly.

After fifty miles, we stopped for gas. Just prior, I had considered asking for a doctor because my stomach pains were so bad. I bought cheese and peanut butter cracker again and washed it down with a Coke. I drank half a bottle of another when the first one took hold, and I left the bottle on the pump. The car soon was on the road, doing exactly ninety. The increasing traffic caused concern for me—Plymouth's ins and outs from the traffic.

The driver related this tale, "When I was younger, I was out on a date, driving on a highway. I had a great car—Chevy SS with a 396.

GARY AND DUKE

Well, we were doing about ninety when this Thunderbird pulled up next to me. So I raced it a bit, and shit did that car go. I was going a hundred and fifteen, but that T-Bird was getting ahead, but then my car got going. When I got to one twenty-five, my front end started rising up, y' know." He looked right at me. I was paying about a quarter mile, and an accident had occurred. The traffic had stopped in both the passing lane and in the breakdown lane, leaving one lane open. But the car we were in just kept on speeding, and he kept on talking while looking at me.

"I was right behind that T-Bird. But my front end rose, two, three feet in the air. We rode together for about ten seconds, and it was really weird. I was totally out of control, swerving right and left. My date was thrown out when the car swerved suddenly to the right, and she was killed instantly when she hit against the pavement. It was her head that she landed upside down sedan with the occupant of the front seat dangling by the seat belt. There was an ambulance occupying the breakdown lane. People were motioning in the only open lane. Cops were pointing at us, corning at ninety miles per hour. You don't believe me, do you? Well, look," he, this automobile's driver, said while continually looking me right in the eye. He lifted a hairpiece that was on top of his head. He showed a four-inch metal strip that was implanted into his skull. The car passed through the lone lane without anyone driving.

CHAPTER 13

Crossing Missouri

The first few hours standing in place hitchhiking, I just plain felt like I was dreaming in Missouri. I was so hungry that I spent minutes willing myself into control. Last night, I was afraid of dry shitting or having my nerves disassociate themselves. I knew I was close to hallucinating. I couldn't think of jokes to tell myself, none. The sun's position meant it was about eleven o'clock. I instinctively looked down the road at the approaching car. I saw Gary pointing for the driver to pull over. I saw the man nod in agreement, and I was fully amazed that the car pulled over.

I got into the rear seat of the car, sitting next to a woman, and I said to Gary, "Oh, come on, neither of us in St. Louis yet!" Only Gary in the front seat was looking at the driver as if to say, "Who is this vegetable?"

"Where you headed?" asked the driver.

"Boston, Massachusetts," I said.

"That's where I'm headed too," enjoined Gary.

"How's about that," said the driver.

"This guy's a weirdo," Gary spoke. "I shouldn't have asked you to stop for him. It was only what I felt sorry for him being on the road. He's been following me since New Mexico."

"But how he knows you're from Boston?"

"Aw shit because he read it off a ticket I got in Arizona. It fell on the floor with him in the back seat like he is now."

"But I thought you said New Mexico."

"Arizona, New Mexico, California, hell, I don't remember. Can't you see I've been roughing it?"

"Yes, sure, Bob."

"He's just been bugging me," Gary said, head straight, voice full of sarcasm in his recitation.

"Then why'd you have him stop then?" he queried.

"Nah, at least I don't think so." Gary turned around and looked oddly at me, which caused us to burst into laughter, me first. "Just nuts," Gary added. The entire car laughed at the comment. Gary kept it up with mock anger, "But I'm not going to put up with it any longer!"

I was amused by it all while fascinated that since then on, this driver of the car just would not listen to any of my words.

Tulsa to St. Louis was four hundred miles. Missouri's filled with some 250 miles of route sixty-six—two lanes. In Joplin, ten miles in, we walked off the highway to a diner maybe four miles inland—one square meal with rolls and milk. Walking back for the highway, we both realized we didn't know what state we were in or even what natural landmark to look for to have a sense of bearing (location). After one car passed, we slept in a hollow.

"Helps cut the noise," Gary said, putting his arm around my shoulder.

"Echoes it," I silenced the small talk.

New day and miles along the day wore hot. We sweated a lot. We walked beside a tranquil pond, never minding a pause for or maybe to swim. It was in the easy nineties. It seemed as if we must reach St. Louis before sunset. A black, virtually new Cadillac picked us up. I ran for the back seat and won. A most friendly seventeen-year-old with brown hair dangling over his forehead pronounced, "Hi, how you doing?"

"Super," Gary exhaled.

"I can see that," teased the lad. Well, I couldn't get a word in edgewise if I wanted to. These two spoke like long-lost friends. I found the driver interesting and estimated a long like for him. He was clean-cut, and his beard was a special look.

"I'll be back in fifteen minutes, be certain that you don't eat anything," the lad said, sending up a cloud of dust as his Cadillac sped away.

"Now why did he make us promise *that*?" I asked Gary, who was already walking toward a log store. It was a dry goods store with a gasoline pump that atop it was a round red globe that showed the gas measure. Dog, were we hungry. We entered the store and carried on a fine conversation with an elderly gentleman, who was later joined by his wife. We two gripsters bought two packs of cheese and cracker and washed it down with a Dr. Pepper.

"Hold out."

"Yep."

"Bad place to be without a ride."

"Sure is."

"Take care."

"Yourselves."

The Cady drove up, and the lad told us, "I told you not to eat!"

I apologized most sincerely then thought that one of us must be crazy because I was almost starving. We apologized. The chap gave Gary a handshake and then placed a brown paper bag, closed at its neck onto the ground. He said, "Here now, promise me you don't touch the bag until out of sight." He left, and I started to pick up the bag, yet he wagged his finger, which upon scrutiny looked like a sign of the cross.

"Maybe there's a stink bomb inside," Gary told me twenty feet away, which caused me to step away from the bag.

Gary's face read pleased as he extracted from the bag two large peanut butter and jelly sandwiched. On the bottom of the bag was a dollar. At a rail post, we chewed our sandwich and counted our blessings. Across the road, four children at their fireworks stood and stared at us, finally setting off a number of firecrackers and some fireworks.

There was little doubt about it. We were stuck. The store proprietors, he told us three times that the countryside ahead had noth-

GARY AND DUKE

ing in it save the three or five families who live there and sure aren't interested in any Easterners coming through.

"Sure ain't interested in meeting any Easterners comin' through." Some kind of luck was going to be necessary. A dollar from the bag, plus ninety cents from Levi's pockets, we spent on a bus we heard was coming through the inside of an hour. We took a thirty-mile or so ride through some of the hilliest countries! A woman with an infant was on board the bus, plus the driver. The road continuously circled around mountains.

I washed in the restroom. I couldn't stare directly at my pockmarked, randomly bearded, sun-scared face. I dry shaved for the second time in my life and made at least a little better impression on the bus driver who had been worried about me and my looks. The woman was smiling away, happy as could be with Gary, while she took looks a bit relieved at the sight of me now a little bit more evenly shaved.

"No, no, no! There are inspectors all along the route, and if you don't pay for Mountain Grove, it's O-U-T for you. But it's not bad. It's only about two miles from where you're getting out that highway is going in your direction. I'm going the back road." The ride was air-conditioned and nice.

The hitchhiking did not improve. We had to split up. "Under the arch, the tallest man-made monument in the world—at noon or midnight." Gary again won the coin flip, and I must walk away further down the road for my own ride, disappointed that I won't be seeing the sun set off the arch. But I was especially pleased that a diaper truck had picked me up and let me sway my body in and out of the sweeping doorway, and its radio played a swinging Dixieland tune just as the sunsets.

The arch had an electric transport to get to the top, but the doorway of the stairs was closed and locked. I waited for Gary while I studied the exhibition.

In the settling of the west innumerable events, important events transpired, so how could one exhibit ever be enough? But inside the arch called the gateway to the west, one stood before a statue of a man depicted in roughshod leather and animal furs gazing westward.

The art played here was a most the man's face out for a moment. The west had been overcome, and I would not want to be of the period shorn by the statue—these times being difficult enough. But the perseverance that was requisite for big adventures held me in personal comparison for a moment.

The gateway arch was 680 feet tall, made of aluminum and silver in color. I toured the ground exhibition, then I went outside to wait for Gary. Yet I saw Gary approaching, and tried as I might to surprise him, I didn't. Inside, we found the door to the stairs was locked, then spending seventy-five cents, we rode in Ferris wheel like cart up to the top of the arch. I quickly deduced that the air-conditioning wasn't working, and the sight in summertime stank. The view was colossal with below the vast Mississippi in her dark brown color with occasional streaks of orange by barges attached together pulling the loads—here, the flatland of Illinois and Missouri. We walked near the stadium when Gary led me toward a bar that inside of it had many a brass instrument lining the walls, while outside was a nickelodeon that here called for any traveler from in or out of state. We inserted a nickel that at first use one each of our eyes! Gary admitted to having ten dollars—a night on the town with thirteen dollars. We walked downtown, reached the city bus terminal where for a few dimes, we could hold our gear that was not particularly large but remained cumbersome for any social event. We followed the hotel crowd, mostly teenage girls, past a handsome, very handsome balustrade to the mezzanine then to the grand ballroom flanked by stately mirrors and a finely polished wooden floor.

Inside the ballroom were maybe two hundred girls and thirty male escorts, the guys who appeared to be uncomfortable in their tuxedos and who all seemed to be wearing glasses. Sidestep momentarily? No, I passed a few couples then went onto the floor where there was an attractive girl, and I, in my clean white T-shirt, asked her to dance. Her name was Mimma, and she had straight hair going three-quarters of the way down her back. We danced to a small musical band while I held her close. She sighed.

"That's the state I'm in too," I said in a soft, deep voice. "You're so attractive. I just can't help myself," I quietly said while holding her more closely.

"What brings you here," I asked her. I thoroughly enjoyed her gown and waist. Mimma was at the state convention for Bluebirds, being held at this hotel tonight.

"How nice. I'm hitchhiking to Cape Cod."

She removed a glove, saying, "Ooh, your face looks so sore and sensitive. You must be hurting."

Mimma and I conversed just like we'd known each other for some time. After perchance five songs, the music ceased playing. We stood nose to nose and just let the ballroom empty.

"Want to go somewhere outside for a drink?"

"No."

"Then do you wish to go downstairs to the cocktail lounge for a drink?"

"The chaperones, they won't let us."

"The chaperones won't notice," I said convincingly.

"That's just it. They would," Mimma said. She pointed to a couple that was staring right at us and said, "Some of them are friends of my parents, and they are keeping an extra careful eye on me."

"I can see why," I said. She smiled.

"Can't you just slip outside then?"

"No."

"Oh, come on!"

"No, I'm sorry, but the chaperones just won't let me. Besides, I'm eighteen in four days." She simply smiled at me.

"Can I come up to your room?"

"Yes." She needed to get her roommate and tell her the news. Gary could be her roommate's date.

Mimma said with bounciness, "Sure!" She continued, "We meet you in room 714. But we have to split because if we all take the elevators, we'll be caught."

I enjoyed the soft kiss she left on my face.

"Like we're guest at the hotel," Gary told me. "Act just like you're a guest at the hotel, and they'll be no problem 't all."

Inside the elevator, my eyes darted from person to person to at last train on the eyes of a business-suited man who was intently looking at me and who stepped inside the left side of the elevator. I exited on the right, leaving Gary the tugging on his arm. Well, in ten seconds, there were three suited men chasing us both down the stairs, and we ended up in the basemen (missing the street by not counting the mezzanine stairs). We ran into the men's lavatory then backtracked (and notice the men thumping it down the stairs). Then we faced an elevator whose handshake was going 3, 2, 1. Before the elevator reached the basement, we were cornered by the guys.

"Leave the hotel, and you won't be charged with breaking the morals of a subadult."

"What are you talking about?" I asked.

"Leave," they said.

We headed for the lobby door. Only I noticed there was an adjacent soda fountain store, and I suddenly wanted a Coke. In five minutes, the store had its revolving door to the street locked. I looked toward the hotel that was sharing a doorway, and there stood three of the hotel's guards and two policemen.

We were separated for questioning. When the cop led me into one room and Gary's led him into another room, what ensued was straight out of a *Keystone Cop* movie. After questioning us, each policeman took each of us into the next room. We found nobody there and did an about-face for the second, third, and fourth room. Every time they turned, Gary and I waved at one another and made faces. This went on repeatedly until all four of us collided as one.

"There they are," my policeman said. "They're just looking for a piece of ass." With that said, the tones of the situation went down a notch or two. My cop took my switchblade away, saying, "This would be better off in a wastepaper basket downtown, ya' say?"

"Sure," I replied.

Gary's policeman frisked Gary and found a slingshot that Gary used to shoot at signs.

"Where are your marbles?" the cop said to Gary. Try as we might, Gary and I could not control ourselves, and we laughed and laughed and had to prop each other up to keep from hitting the floor.

It was not possible to keep the cops from thinking we're a couple of wisenheimers. The policeman said, "If you broke and don't know anybody in St. Louis, I'm sure we can find a place for you. I'm sure we can work something out." We were taken to a prison, a large place.

With paperwork going on between Gary and his cop, I sat down at an adjacent desk. I put a goodly number of paper clips into my jacket to put together in a cell for something to do. In another room, we were told to take out our possessions. It was now that I realize I had a bag of grass inside my coat pocket's lining. My amigo Gary was looked over with a fine comb. I decided to stand and hold my jacket by the reversed pocket and dangle the coat to show both pockets are empty.

"Where else do you have?" the officer inquired. I pointed to the top of the desk where I had amassed a pile of perhaps a hundred paper clips. The cop winced. Gary laughed aloud. Fingerprinted, weights and heights recorded, we were taken in an elevator cage or crib left and right and up in prison.

"Don't look in at the guys if you know what's good for you," the cop said plainly. Bars were drawn up, and other sideways went. Then as we exited to one of the main floors, the bars were locked most solidly.

"Hey! What's our charge?"

"Loitering or trespassing on private property, take your pick."

"We going to face a judge?"

"A judge, ha." The cop laughed.

"How long we in for?"

"Twenty-four to as long as you'd like to make it. All up to you guys. So watch it."

There were no blankets, no pillow. There was an open window opposite my bench, and cold rain came in on me for the night (until what must be 3:00 a.m.). At what must be 6:00 a.m., a guard with a club brought each incarcerated person a day-old doughnut and a hot cup of coffee. Gary put his down, wrapped his finger around the cell bars, and asked for another doughnut. Instantaneously, Gary was looking closely at his clenched fist as the guard had hard and fast struck his club against the bar in record time. Maybe at nine o'clock, we were let out of our cells and directed into a single row.

"I really don't want to spend the day popping rocks with a sledgehammer," I said to Gary. The yard had a sign, "No talking."

"You two c'mere," a cop with a clipboard said after checking off the rest of the line. I recognized him as one of the guards who arrested us last night. "We don't need you today." He looked tired and motioned to a guard behind glass, holding a shotgun who threw a toggle switch, and a world of bars had one row open. He did look tired and said routinely and flatly but covered the bases. "Get out of town. Today, before noon."

"Hey, Tom," said another policeman. Going over a clipboard, he said, "Do you know that you didn't check the one out for possessions?"

"Jez, I was tired even then. Gotta skip the coffee and go home." I gulped.

He said, "These double shifts are killing me."

"Let's get our gear," I said. "And go back to the hotel for the girls."

"Or go to San Simeon," Gary tartly fired back. I looked at Gary and saw his hat was dusty. And his pants were worn thin.

"Hey! Why don't you talk?" I declared to Gary, who backhanded whacked me in the throat in response. I wanted to counter with my right fist, but stomach pain changed all plans. Seeing my reaction as if on clockwork, Gary had a stomach convulsion. We could rather beat one another than go through this.

"Hell, we're both beats," Gary broke the ice with and continued to say, "Let's move."

"I suggest we take a bus across the Mississippi...you see how wide that is," I said.

On the bridge, at the opposite shore, the city of East St. Louis, it appeared even more congested. This span was quite lengthy, as one would have it so is the river while (one would guess at this time) there was about to be built another bridge crossing the Mississippi. On the opposite side, there were hundreds of girders than busy crows in a great cornfield. Cars sped past us as we walked across the Mississippi

(Lord, my Lord, I love you). There was no hope of getting a lift because there was no breakdown lane. We walked on the wire mesh that separated the bridge's directions.

There were miles of bridges to go (Lord, help me). I feared the worst five-mile stretch possible with only the bridge and the river to be seen. But wait, a 1950s gray sedan pulling a multi-wheeled cart stopped just after the single bridge ceased. It would be dicey riding on the rear cart that had no containing walls, but what could you do? But wait, we were waved inside the car by an elderly man with a silver beard. Accompanying him was a lad who appeared to be his grandson.

"Sure looks like you could use a ride," the gentleman said.

"Sure could." I smiled.

In five miles, still, where a bird's eye view of the Mississippi is with a mile of no breakdown lane ahead, we were let out at an off-ramp. We split up, and Gary won the coin flip, so I must foot it.

"A few miles back, a gas tanker flipped over and exploded. You must have just missed it," the next driver who picked me up said.

We drove right into Indianapolis at exactly eleven thirty at night. We let out at a busy on-ramp. Then tooting away was that all-American family indicating to me to "come one!"

So I had six hours to kill before meeting Gary at the city's famous central fountain. What to do? With something like four dollars to my name, I went to a double feature of Clint Eastward movies, *The Good, the Bad, and the Ugly* and *A Fistful of Dollars*. At first, the air-conditioning brought a fine form of contentment but not after a couple of hours. A half-full box of popcorn found underneath one seat was tasted some then discarded.

By one-thirty, only tourists were still milling around the fountain. There I located Gary. By a number of military field pieces, we played soldier. Then we noticed a Laundromat. Gary said we needed to go there to spiff up for the final push home. "And waste our money," I said, trying to conceal that I had been at the movies.

"Hey, kid, sorry to tell you this, but I never could understand why that girl at the hotel found you attractive because you look grubby. I'd make you take a shower before you."

"Hey! Screw you!"

We located an all-night pharmacy and bought some razor blades and some detergent. This store doubled as a grocery, and Gary wanted to spring for a watermelon.

"I'd give twenty cents," I said, "for what a watermelon is worth." Gary dug in his Levis's pocket to produce two dimes. Gary shared gobs and gobs of watermelon, and he looked bewildered when I handed him back two dimes. We laughed and tossed pieces of the fruit back and forth at each other. Our bellies ached, and the watermelon was three quarters gone.

CHAPTER 14

No Soap

Warm, dry clothes were repacked and put on. It felt good to feel warm in just washed clothes. However, said Gary, "It's a good thing it's nighttime. You should see your reflection in a mirror." I went to a mirror and saw red hairs popping out all over a good part of my face. My arms had bites, bumps, bruises, and blotches. We walked several blocks, taking in the fact that it was a commercial area, and buildings predominated. At one stop, Gary drooled over a switchblade knife, saying that we'd best stay there until morning to find out its cost. We had very little punch toward the eight cars that all passed in a row. At the eighth and final car, it stopped.

"Where you guys headed?"

"Boston."

"You headed in the right direction?"

"Boston, M."

"My! Have you got ways to go? Tell ya' what. I can cut by an old field I know and save you about ten miles… because it bisects the way the highway to New York goes."

"Not according to the map," I told Gary in the front seat. I grew suspicious.

"Nah, the map must be old," the driver told us. "Must be fifteen miles it saves." The man looked in the rear at me and smiled, and he added, "Sounds like you don't trust me."

"Just doesn't seem to fit," I responded to the driver.

"I'm just going home. Believe me. I grew up here 'n been back here since I was in the service."

I sized him up good. He was a large man, about 6'3" and 240 pounds.

"I bet you meet a lot of weirdos going as far as you have," the driver spoke. Gary and I did not respond to him.

As the car traveled about fifty, I noticed the rear seat had no door handles. I studied the man closely, looking for any sign of weakness. "Of course, you have to look pout for people trying to butt fuck you. 'Course it's not so bad if you just pay them back for helping you out with a BJ or something—courtesy of the road…paying a man back for helping you out. I bet you two wouldn't have gotten half as far as you've gotten without helping being friendly." I got attuned to Gary shrugging his shoulders and what that meant. The car was ever so quiet, except for the man doing the talking. "That's not too bad. What are you, nineteen?" the man went on, oblivious to the silence around him. "Deal you need. I used to do that myself when I was your age."

"Ahhhh, you're a fagot," Gary emphatically said. That amused me while I wondered where in heavens name we were situated, in what state.

"A fagot! Hell, I was in the service buster. It's all right."

"Ahh, you can kiss your photograph. You're all alike," Gary spoke sharply.

"I'm stopping the car right here and now. And you are more miles away from the highway than where I picked you up. Now you do what I'm telling you, and you just might reach the highway. It's four in the morning, and there's no police around. They're all sleeping. I know I'm from this area. Now I don't want to intimidate you, but I know the martial arts, and there are no door handles to get you out. You might not get to see the morning in one piece." He took off his glasses and started to unzip his trousers.

"Now look you," Gary started to say with slow-paced words.

"I'm not looking for any pretty words," the driver said, cutting Gary off. "And don't you try pulling a knife or anything. I see you back there. I'm a trained professional soldier. Now take off your clothes!"

"For cripes sake, mister!" I blurted out loudly. "It's against our religion. Do you understand what I'm saying?" I tested the two-inch knife blasé at its point.

"What's that … your religion. Oh gosh. Oh, gosh, I'm sorry." The car turns, and we return in the backways about seven miles, in about five minutes because we go so fast.

"Don't mind the speed. I know where we're at. I'm from around here. I just want to make it up to you guys."

We came to a stop at the anon ramp to the highway and were let out. I exited the car lastly. When the man returned and began to drive away with the rear door open, he said, "Just what religion is it that you guys practice?"

I made certain that we're well out of the car and slammed the door, saying in an attempt to imitate, "Ahh, you're a fagot. Forget it!"

CHAPTER 15

Echo Chamber

Metallic echo chamber like walked down the darkest on-ramp. The almost incessant bird chirp ceased now. I read that my body was in trouble just trying to walk down this pavement, this one maneuver.

"I'll be an SOB if I don't make it," I uttered. Gary stopped three paces further ahead. Time sounded in. Our knees were rocking north and south. This special no-sun time had us awake in a stupor. In this open expanse, surrounded by dry summer grass and the present road, I tried to think. I opened my eyes every forty-five seconds. I had to rest and reconnoiter with the physical world. I tried to think. I stopped from gas pains, this time just trying to move the right leg forward. I drew a short breath, knowing Gary won't be coming at me.

"Gas," I spoke. Gary simply nodded. We were not untracked, only concerned seeing one thousand miles of road ahead.

"I can't sleep, or I won't make it."

"Yeah," Gary agreed. "I know what you mean."

"Just because I'm suffering from something and even if it is malnutrition…is just the way it is."

A limousine speeding past snarled thick heavy tires against a sidewalk part of the road, and a horn called us to turn around. There in the dark in the darkest night, a car piercing the night with cardinal red taillights stopped in the sphere of darkness. Commencing to run

GARY AND DUKE

and then slowing to the pace of mouth-old coffee, I approached one of four handles. The car was cavernous.

"Hello gratefully," I said to the driver, producing a full toothy smile. Gary approached the car and hid in from the windshield. He made a face that the driver and I laugh at. I was still taking in the interior of this car when I heard the drive spoke, "I'm going to New York City. Delivering this baby to an airport on Long Island." He asked where we are headed, and when one of us said, "Boston," this man said, "I'm sorry, I'm not going that far. That's a long way from New York City." And he meant it! This large, shouldered, smiling man was apologizing for going a thousand miles.

"I used to drive trucks and cars across the country," he said. "Only the big ones or the limos now."

"Must be nice," Gary said. "I mean the scenery."

"Not really. Most of it is just roadwork. 'Course if you want to, you can go off the road and see lots of beautiful places to see." His name was Bill. He spoke a bit defensively then mentioned, "There's plenty of beautiful sights." The ride became quiet after five miles, which was typical. I tried to keep the conversation going in that way, paying back for the ride. But Bill was an experienced older man who knew better, and he directed us to sleep. Upon his third attempt to get us to rest, Bill was shouting, "Get into your sleeping bags. Unroll the four rows of seats and get into your bags."

"What's the matter?" asked Bill.

"I was just studying your face," I said.

I saw Gary like myself weaving on our knees, trying to enter each of our sleeping bags, fingering and forgetting and anything but laughing. Two tires pupped fingering and forgetting how to open a sleeping bag.

"Four rows I said!" Bill shouted. I told you there were four rows of seats to undo. "Now let me drive, and you get some sleep!"

Later on, I asked, "How long have I been asleep?"

"About an hour," Bill spoke, tempering his words. In what feel like three hours, I ground my eyes open and saw Gary sleeping and did likewise. Quite a few hours went by, and after chastising both of us for snoring, Bill took the car off the road only to, later on, cause me to shout at him for his snoring.

I looked at Gary's face and studied him, then being caught, I played the fool until I caught myself being one.

"We could stop in Pittsburgh, take a good long sleep, and meet some never met relatives of mine," I said. "And we could get some food."

"Talk to me about it later," Gary said face-to-face.

Overhearing us, Bills asked, "How about some pancakes and hot coffee on me? With sausage." I laughed at his phrase modulation. The second serving of pancakes and the second pot of coffee brought some life into us. Bill stated that Robert F. Kennedy had been killed in California. He repeated his words. Bill spoke of his wife, who had died three years ago. It was a quiet moment, a pink moment.

"Pittsburgh is in about forty-five miles," Bill shouted to we awakened boys who found inside half a mile a sign, "Pittsburgh forty-five miles." The breakfast nutrients allowed us to travel much more comfortably, traveling fifty-five, lying down inside a cavernous auto. Pittsburgh rose through a cloud of industrial soot. Nature was indeed up with soot for miles around. The country of the land was still green by far, yet here, a few miles further was another variation of green. Apart from the sculpture of the major cities, nature reigned. We were tired at a look, but with another look, we'd make it home damn Pittsburgh or any other obstacle.

Inside the car, reading a road map, I realized that in a few miles, we would reach Johnstown, Pennsylvania. I looked outside at the environment because I was reminded of my uncle Paul Fitzsommons who wrote of the flood of 1989, not the flood of 1936.

The South Fork Gun Club had upon its premises in 1889, a damn blocking the upper flow from a lake used for sport hunting and fishing. Ten days of rain made the dams level precipitous. People from the town spoke to representatives of the club. They spoke to get the dam's level to be lowered. It was not done. The dam broke, and devastation ensued. Four towns were hit. Johnstown was hit the worst. Trees factory pieced. Chemicals were all sept up and away. When burned by bridges, the newly created river flowed backed up.

Twenty-two hundred lives were lost. One could still see ravaged earth where a temporary river existed.

"Stopping at Pittsburgh will spoil our doing the trip solo."

"Your sister," I said. "For cripes sake, we had the bus ride and JMcK."

"Oh, don't remind me of that ___," said Gary.

"Okay," I responded. "But the rest of the way won't be easy."

When Alentown and Bethlehem, with their tremendous steel mills and populations greater than one hundred thousand, were seen, I had earlier longed for a sight of the population save the never-ending road.

New Jersey had some wonderful green spaces, and the limo was going at ninety-five.

It was not too long before Newark was arrived at, and gray was lining the tops of ornaments on the building as we entered the city. Really the powerful dark shadow of New York City came from lights piercing a fog's mist. Bill had earlier floored the limousine's gas pedal, kept the car at an even seventy miles in an hour. It was New York City that was of optimum interest now to compare with the land, people, and the rivers. Newark was duly noted. Then Bill said he could drop us off in one-two places, which we did want. I rummaged anxiously for the map, growing distraught, finding no section blown up for New York alone. Bill had a look in his eyes that was genuinely glad to be helping us out. He had been so kind to us.

"Take care, you two. Don't look so hot," said two worn blokes, now to be inside the city our paws on the mat. The pavement was good and soiled beneath the boots, but our legs and constitution were weak. Blocks, blocks, blocks of "exhausted" lived New York drab, faded gray.

"This is where all the people are," I spoke, not trying to hide the fact here is where I would soon be going to college in the fall. We stopped and stared.

Gary curled his finger on his jacket, saying, "This is where you'll be going to college. I don't know what Kansas will be like, but I wouldn't like to come back here." I didn't let him dominate the point.

Instead, I took in the angles of the sidewalks and buildings, feeling somehow attracted to it because I was to be a part of it.

Gary asked a streetwalker for directions. We were to walk three miles. "No, there ain't any breakdown lanes. They are to come to a bridge. Somebody might be nuts enough to stop for you. There's a lot of weird people in this city." We made to leave when she said, "Where you going, Maines?"

"No, Massachusetts."

"Where you from?"

"Boston," Gary evenly replied.

She couldn't get it. She was open-eyed. "I thought Boston was like Noo Yawk. That guy must be with you. Nobody looks like that around here, mister," she spoke with a question in her tone, trying to pry an answer.

Six-thirty in the afternoon and we walked off street corners watching shades brace each corner. Upon walking two and a half miles, we got to the bridge where crossing on foot was impossible, much less any car would stop—daring danger as it were. Two-twenty cent bus rides got us out of Brooklyn, over the Bronx-Whitestone Bridge near North Pelham in the Bronx. The bus ride was not at all what we expected. Gary sat to the rear right of the bus, I on the left. Three seats ahead, there was a hippie girl, denim pants, straight brown hair wrapped by an Indian scarf, about nineteen-looking at the age when teens are leaving and acid has left. For a long time, she smiled at me, and I knew something you don't smile.

Gary got up and moved. She sat down in front of him. But she got nowhere with him and pulled on the brim to my Australian hat.

"Got the combination to a lockbox?" she asked.

"I got mine," I said dryly, most blandly looking into her eyes. At the next stop, the three of us exited the bus. The girl stood right next to me as if reunited. "Where you two heading?" she broke the ice with.

"Boston," I said, avoiding her downturned head and up pretty timing eyes.

"I thought that's where you're from." Her bandana and hair both would look a lot nicer if washed. "You should have stayed on that bus," she said at the accelerating away bus. "Now you'll have to pay another fare," she spoke, looking hopefully into my eyes expectantly.

"We want route twenty-two," I deferred.

"No, no, no! That's the last bus you'd want because it goes through every town between here and Connecticut. Your best bet and I doubt you'll get a ride tonight because of the way you two look—" she paused to let her words register—"is to go four miles north to the highway-situated inland." She shrugged her shoulders, indicating the less than fortuitous situation.

"Forty cents?" I asked her. Brute Force got piqued, taking half a step forward, then stopped his approach to fold his arms and watch.

"I haven't got it," she said. "I was going to ask you." She said, "You can spend the night at my place, and I'm not going to say it again. I've got some mescaline, but only you can come." She stayed with us while we waited these minutes for the next bus. Gary had to pee, and she went with him. Gary returned first as he aggressively pulled at his zipper. After that, he rubbed a sore spot on his arm. I'd decided that I couldn't leave Gary at this point in the trip. We skipped the bus and hiked the four miles, but first, I must unclench my fists for my knee sockets have locked then slipped. It was painful. The road was as dark as a closet. We walked on a very elevated sidewalk. It was impossible for any car to stop here.

When we got within sight of a diner, "Let's get going!" Gary said, loudly enough for a passenger in a passing car to hear. This time, Gary froze in stomach pain. He extracted his wallet slowly and pulled out the two well-pressed but soiled two-dollar bills that my mother had given for emergency. Gary said he had been shooting craps. Inside the diner, I eyed the menu as I've never done before, as the waitress lets me take my time. Gary ordered a plate of roast beef, mashed potatoes, and mixed vegetables. Awkwardly, I asked Gary if I could have the same thing.

Gary said, "Sure but don't go ordering any lemon meringue pie." The waitress and Gary got along famously, while I was reticent and brought them to both say, "He can't talk."

On the way out, a policeman held the door open for us, and I came to think, *New York has people just like at home.*

To sleep or hike wasn't particularly opportune as the area we had come to was developed. So we'd walk to get to the familiar inverted vee of a highway where on-ramp met highway and stood by two four-foot pillars.

A Volkswagen bug came to a stop there, right inside the small vee, a bug with Pennsylvania license plates. The car had a mid-thirties couple inside, friendly, inspirited in verse, and smiling. "Hey, man, how ya doing? Where ya from?" asked the bearded man through even lips, but he got no reply. Gary and I each wanting the other one to talk.

"Boston," Gary said.

"That's what I thought. Well, we're going as far as Norwalk to be visiting some friends. Want to come with us on the coastal route?" The fellow was most relaxed with shoulder-length brown hair and particularly sensitive to our having hiked mucho miles. Both of us partners appreciated the woman who smiled and said, "We're from a hilltop in Pennsylvania, the far south, and have a house on a mountain top there. Getting away from it for a while…too quiet there."

"Pennsylvania is beautiful. Must be very nice to live there?"

"It is," he said. He went on, asking, "Do you have any pot for sale?"

"I got some for my brother. In California."

"I can dig that. What did you get, a lid?"

"I don't know what a lid is."

"Four fingers by three fingers in a plastic bag could be an ounce. A lid three fingers by three fingers."

"Well, whatever. I got it for my brother."

"How much did it cost?" It had been drizzling, while now it was a rain that dotted the windshield. I was comfortable, secured in a warm bug going twenty miles outside of New York City. I opted to travel along to Norwalk, for my map showed it closer to Boston.

"The rain isn't going to stop," the driver said, fiddling the car wheel. "Says on the radio that it's a hurricane." He laughed aloud, bit became more to them moment interest, the talk becoming more animated and involved for fifteen minutes. The man said he'd hitchhiked the container twice. "Or was it three times?" We were all sorry when the car had to stop. The boys got out and hit an open pouring sky as a storm approached.

Left at a truck stop/diner, Gary wanted a coffee. He discovered in the breezeway that he'd lost a part of the remaining money. We waited an hour with it still pouring. It was getting late, so we returned to the highway's vee. Replying overhead visibility lights from the truck shop cast a sickly yellow, red, white silver. Trucks roared past. We stood in place. Over by where the trucks parked by the diner was a twenty-two-year-old long-haired with his girlfriend. His arm was around the lass, but it did no good for the rain coming down on all of us. I extracted from my pack that horse blanket that I'd taken in California, and holding it with my grimy hand, I thought of how I wanted to hang it on the wall at home.

"Here, you, this is for her." I tapped the fellow on the shoulder with a friendly line. "Okay?"

"What's your address? I'll send it back to you in the post."

"No. Just keep it. It's from California." Again, the peace sign was flashed at us. We, buffoons, started a little reverie by punching one another.

"Notice the weather, Gary?"

"Mmmmm. Falls's closing in." We were trapped by circumstance just in a storm cold after being in a desert a while back. I thought July was almost mid-September.

Fortunately, we could not get more soaked than we already were. In the pour, we stood immobile, listening to the purrs off the highway. The slants of rain went this way and that.

"Look at your stomach!" Gary shouted, pointing to my waist. Hands on hip, I saw my stomach in between hands, large distended and bulbous.

"I'm almost starving, Gary. This is impossible!" While I undoubtedly had lost several pounds on my 170 frame, I looked like I put on pork chops by the dozen, a red lobster in the face, feeling steam rising.

CHAPTER 16

Nearly Finished

It was one thirty. We were soaked. We got a lift from a teenager returning from a date in Newark. I smelled alcohol on the teenager's breath. The fellow was all impressed with what Gary and I were doing. He prodded me with many questions.

"It's cool," he said. "But I don't think I'd do it." After thirteen miles at a two-plus time, I stopped watching that curios item of a clock. And the guy let us out, saying, "Hope I didn't scare you going that fast, but the windshield wipers don't work, and I'm out running the storm. Radio says she's a real doozie, so get some shelter. Get out of the rain though. I really wish you luck." This last remark caused me to burst out in laughter, and the kid looked at me like I'm crazy. But that was all right. The kid drove right away an off-ramp, leaving us the road.

Maybe this was in Bridgeport, but it was a funny spot as we'd ever been in. Curiously, the miles of road ahead were a hundred feet up in the air while supported by a very, very long super system of concrete and pillars. It ran for miles ahead. There was a river below. There was no emergency lane. We were waiting in a storm's lull on a highway that would only bring early morning traffic going to work. And what odds would this bring? We laughed.

First, I should mention that this whole elaborated system of elevated roadway was matched by a similar sister system of raising paved roadway in the opposite direction—bridge way pointing south. Trucks in their own gushes of water inside the storm sped away from their wake, creating a blast of warm water that hit us like some flock of birds flush in the face. We turned toward the upper ground of this enormous, slanted roadway. When I turned backward to see an approaching truck, two puddles hit my face flush. I checked for blood. I really did as it was a sizeable jar. Gary, standing but feet away, couldn't hear what I was muttering.

By the exiting ramp were two five-foot stone pillars. I crouched behind one, and Gary bent down beside his pillar. This proved good relief as the storm came on hard. Occasionally, I looked out at two o'clock point to see well off in the distance lights of the coastal way shine through the rain and wind. Some of the messages of God came to me. Forcing an earache, the warm air barometer pressure snapped my eardrum and sent a shock wave all but too intense through me.

"For four minutes, I must not think," I said to myself.

"Wow! That was really something!" Gary cracked. I stayed quiet. We hadn't been communicating anymore. I tried hard to crack jokes but gave up in the first few words. We could hardly complain or ask for help.

I decided to relax and entertain my strained head. I saw Gene, saw his curious and then cute face trying to figure me out. Then I stepped away from him and went so deeply back in the dream…the wraith.

Again, I had been stabbed by a beefy, long dark bearded man, the 1500s, drunk, lordly pirate with red eyes, toothy smile. He had earlier terrorized me or attempted to by insulting me before the table and people where we stood, throwing pewter ale and wine mugs at my head. He threatened to throw one straight into my face. The crowd to a fiendish frenzy, the crowd began to vociferate me to be put on my place. He had to prove that he was lord, and he would kill me if he had to display his sovereign powers, or I would be parried. I envisioned the blade. I was almost a hundred. I said something but could not hear it, so I said it again louder. My head hurt, and I was in pain. I realized I was 175 miles from home (wham). That I realize where I was (whoop shaloop) with Gary shouting at me. Another blow coming from nowhere was coming straight to my face!

Cringing, I saw Gary's palm go past my eyes. There was a strange delay and then wham boxcar in the ear.

"All right! All right!" I yelled. Then I got it again. "All right!" I screamed so fully that it hurt.

For six minutes, I had to tell Gary that I was all right.

Boy! Were you gone? You should have seen the look on your face. A very large factory clock showed the time as four thirty in the morning. A wooden trailer block with an inch head nail was some ten yards down the highway road. I picked it up and began toying with it, tossing it up in the air. Only I became preoccupied with the block. I beamed from the toy's magical propensities. Then loosely, playfully, Gary knocked the block from my hands and began to soccer the block of wood. This was a loving, thoughtful move. We played, and this wore us down.

"We gotta cut this stuff out!" We saw birds on a railing and tried to decide which of them would be best to eat. We went to a higher elevation on the bridge when the rain stopped and stared at the surging bank of the stream/river below. There was a drizzle left, but it was just moisture in the air, spewing by high gusts. And we looked for the approach of any vehicle. There were many miles to get off the highway. Finally, we found from a quarter mile away the lights of an approaching car shines.

"Hey, we've got to get this car to stop," Gary said to me. "Pull out your bag of tricks, and see what you can do."

"How 'bout if I put one arm inside my shirt and act like I've only got one arm?"

"Nah, that'd scare the piss outta them."

"Act like I'm dead?"

"Nope."

"Like I'm injured."

"Yeah, that'll do it. And I'll come running up like I was going for help. That way, they'll just see you at first and maybe stop."

GARY AND DUKE

The Connecticut state trooper's car stopped about five feet from my prostrate body. Gary came running over, saying, "We walked across New York City and are broke."

"Get in," the trooper said pleasantly enough. He had to tell Gary he was not bring arrested before Gary gets in the car.

"Ah...ah...I've got this friend with me too."

I approached the car, wiping my brow, saying, "Is this the state flood relief?" The trooper softly conveyed that he was glad to help us out, which made him feel good. The blow from the car's heater was felt in warm swirls.

"You didn't have to lie down in the road. I would have stopped. Boy, that was some storm, wasn't it? Nobody could have been out during that one. You'd have been stuck out there a long time. All the roads are flooded. What'd ya do? Just get out of a car after the storm broke?" We boys tugged at our mustaches. The older man tickled his own mustaches, saying, "The hiking is better a few miles ahead when these bridges stop. Some place for some kid to hitch from though, hungh?" There was no reply.

"I can see just by looking at you that you know better than to hike from one of the side roads. But that storm makes one do strange things." He asked where we are from. We offered to show our driver's licenses, but he said, "There's no need. I can see you've been traveling. And you're not on the lam, are you?" He was kidding. "You should avoid the highway and take a break from the weather." The trooper was quite relaxed and friendly. "The storm is over, but it's going to remain pretty miserable until daylight. I see you have your sleeping bags, and I'm sure you have ponchos. You ought to sleep off the road. You're not supposed to hitchhike, but I think it's okay." We still said nothing.

Breaking the silence, I spoke, "We gotta get to Boston."

"But it really isn't that far," the man continued cordially while intently said, "I suggest you get a couple of hours of sleep, then later in the day, they'll be some traffic."

"Yeah," Brute said sheepishly.

The trooper eyed his wristwatch and said, "You can stay as long as you want in a cell. I can promise you a pillow and a cup of coffee and a doughnut." My stomach bifurcated at the notion of a coffee.

"Gotta get to Boston," I said mechanically, inflexibly.

"I swear you'll get there today," the policeman said, just maybe thinking we might not make it without rest. Gary and I exited the vehicle, and we stood there, looking at the car and at the low-flying dark clouds penetrating the moment. Within a few hundred yards, we arrived at the base of a tall hill. A car passed. Another car passed. The third car was going about eighty stops.

The long dark car with taillights so far its headlights was a limousine. Into a green limo with a black interior, we went. But before getting inside, I checked to see that it wasn't the same car. Then at the front handle, "Hi!" I said, full of spice.

"Hi!" an open-eyed teen with black horned rimmed glasses said. "I bet you never got a ride from a limousine before!"

"Actually, we got out of one a few hours ago after a nine-hundred-mile ride," Gary said from the third row of seats, arm dangling over the rear seat.

"I'm going to Wallingford, just got this baby from an airport in Queens, New York." Gary and I looked at each other. It was still not the same car.

"If that's toward Boston, that's just fine," I wheezed out. I stared blankly at the dashboard clock in a semi daze. The car's vents were shut.

"I've got the vents closed," the kid said, "because a while back, the windshield wipers froze, and the engine started knocking. Sorry, I'm not going that far [he said at ninety miles an hour], but I gotta outrun this storm."

"There's more?" Gary asked.

"Who knows," the toe-fisted driver said, "it's a hurricane, and they're unpredictable."

After a couple of slides on the slick highway and a half an hour, we were let out. Then we stood inside a drizzle. The day's overcast showed little light. I remembered that the last several rides had told us to get some sleep.

"Damn," I said. "Let's get some rest over there," I said, pointing to the breakdown lane where there, on the right, were a couple of road construction machines. It began to rain fairly hard. After a moment, Gary suggested we retire underneath the same machines.

"What if they start up the machine while we're sleeping?" I asked.

"I guarantee you a thousand bucks nobody is building a road today."

"A thousand bucks."

For hours, we slept beneath a low-bellied road surfacing machine. Awakened, I saw a roof inches above my head, and Gary was sleeping. I got to sleep, and in a while, the procedure was repeated. This day we never saw the sun, while when it hit its zenith, we slept inside sweat-filled sleeping bags and said nothing.

"C'mon, let's go," Gary said near the end of the day. We stood there, thumbing until the moon showed it to be about nine o'clock. It rained for half an hour before we gave up, looking at the last cars with malice.

I couldn't comprehend why so many cars passing the next morning had people all dressed up with ties, suits, and dresses. And they were smiling. A holiday? After several hours of standing, thumbing, I saw Brute Force, looking at the approaching cars with a glare. I patted his back, and it boosted Gary immeasurably.

"Hey, I got it!" Gary yelled in the night. "It's Sunday!"

"We're looking at the car too intently," I observed. "We should be looking beyond them."

"It's pitch blackout, you idjit."

"They can see us."

"If they're going to stop," Gary said, "they're going to stop. They can see us like a stop sign. We're the only thing on the road that's different." The nightcap was pulled down, and the overhanging clouds made it markedly too dark. We went under the machines for another night, ready to swear in the first magnitude. The vehicle that we lie under was rust orange.

We awakened to a kind sun hidden behind clouds. I said within the gray day, "Shall we not rise, fair knight, and slay the day ride dragon?"

"No, let's not. Let's stay put and see if it stops raining!" Gary spoke, emphatic enough to deflect a rock.

"I'm not," I replied. "It's not going to rain for three days with me standing in the same place. I just won't do it. See you later." Brute, like a bear, rose, looking ready to fight. He surprised me by wearing a smile.

"Hey, you're right!" I said, noticing that Gary had pain in his stance. We paced four miles, half of the way, with our back to traffic in the rain.

"I'm not going to look at those cars anymore," I said, one heading off on the off-ramp. "You can see them with their mugs of coffee going for the office. They don't give a damn about us." I pointed toward a telephone pole that had on it a silhouette of a telephone with an arrow pointing. I damned the distance to get even near it and damned any defiance my friend might have.

"You're calling home?"

"Damn right," I said. "I have h-a-d it." I called home, and my dad said that he couldn't believe we've made it across the country and are stuck a hundred miles from home. I stared at the telephone receiver instead of listening in.

"If you're stuck at the end of the day after I work, I'll come pick you up," declared my father.

<<missing page 101–105>>

"Let's go" was all I said.

Gary looked at me but said nothing. Gary made to throw away his money, saying, "What good is it." I corrected him.

The view to the northway had another stone pass overhanging road, facing a forty-five-degree hill. We agreed to thumb for an hour and then took the hill.

"The sun's shining for the first time in four days," I said tremulously, looking as the flow of traffic from the highway increased. Brute looked at me meanly then broke into a smile and laughed. From an on-ramp, fifteen minutes after the sun started showing, 1944 Ford Coupe—raises engine sticking out of the hood, dual four-inch pipes, and a rumble from the finely tuned engine—pulled over on the ramp, stopping traffic. Cars honked. The driver flicked the fickle finger at the other cars and waved toward us to come on. Once inside the car, jerks forwarded to thirty-five in first, fifty-five in second, and drove to a sliding stop of dirt and gravel by the side of the main road. Two feet behind the car was a dime.

"Let's get those bags in the trunk," he said, and all three of us stood by the trunk behind the car. And reaching down for another bag, he said, "I can take you as far as Boston, but that depends on your money."

"We're trying to get ten miles west of Boston to Wellesley," Gary replied. "How much is it you want anyway?"

The driver said, "Five dollars apiece." Well, to tell the truth, he saw my reaction to this and demurred, "Four dollars. Well, for heaven's sake, how much money do you have?" His figure begged for a reply.

"I have four twenty," Gary said, taking me unaware. "Savage, how much money you have?" The man cringed at the name. Looking at the man, I tried to convey that I really was broke.

I said, "I have no money." Yet I knew that it won't hold any water that I was telling the truth.

"Sure. Sure. Sure," the man excitedly replied. "I can see you've had a rough time. Jez, that's okay. C'mon, let's get in the car. I'll drive." The last remark made Gary and I laugh.

CHAPTER 17

Out-Duked

Gary said, "We're trying to get ten miles west Boston to Wellesley. How much do you want?"

"That's almost a hundred miles. Five dollars. Apiece." Well, tried as I might, I couldn't convince him I was broke.

"I have four twenty," Gary spoke with an open wallet. The man was cringing as Gary said, "Savage, don't you have money?"

"No," I spoke.

The driver said, "Sure. I can see you've had a rough time. Let's get back in the car. I'll drive." Gary and I laughed.

Well, the road map from Bill's limo showed that yes, the best way to go was the way this driver said. Twenty-eight miles to Hartford and then headed east for Boston. We got to five miles past Hartford, and the car was turned off the highway. I thank God for all he had shown me.

"I've done this before for a few more bucks," the man said. "But I've got to stop by my house and let my wife know what we're doing." It crossed my thoughts that he might be taking us to a police station, but that was fine if it was closer to Boston. We got to a long driveway, went the length of it. And a smiling woman coming out of onto a house porch greeted us most cordially. We entered the house. She offered us a coffee cake. I had a slice, and Gary, he took piece after piece, eating three quarter of the entire pie while pretending to be listening.

"Do you mind if I see some identification, driver's license, or something," he said then relaxed noticeably.

Back in the car, me in the rear set, I folded the road map into two parts—Massachusetts in thirty-five miles and them Framingham right closer to home. I threw the map out the left rear window. I thought of Bill and the most remote chance we might ever see him again (in heaven?). Passing Douglas, Massachusetts, I wondered aloud, and the gracious driver gave me another road map. Then came Framingham, and I relaxed, knowing I could hoof it home. Looking at the nape of Gary's head, I tried to say something, but my words were gibberish.

After a considerable amount of time, the driver, who kneaded the wheel, spoke, "Hope I didn't mind my asking for your IDs."

"SOP," Gary said.

"You don't owe me anything. I can see you've been through a lot and deserve a fine end to your trip, and you should arrive with at least of the money you left with."

It was apparent to me that the man thought we are going to steal his car. Gary laughed a full laugh. The driver let loose and had a pretty good laugh. Gary handed over his money.

"Here, keep it. It's yours."

We drove past Pond road, the boundary line for the local woman's college. Gary said, "You can let us out in a mile. In the center."

Cutting through the driveway off Church Street, by the graveyard, and onto Center Street, we saw the road considerably covered in sand. A passerby explained, "An eighteen-wheel oil truck flipped over this morning." As we crossed Center Street, a high school classmate was seen approaching.

Splitting up, Gary and I each spoke, "See you later." I cut along the grass by the town hall, walked along the railroad, tracked a hundred or so yards, then came out onto Linden Street. I saw a gentleman, who when I was in school, used to greet me every day when he went for his daily newspaper. I explained where I've gone, and he stood in place, watching me even once I leave. Next, in my road,

Clifton road, I saw Mrs. Dorothy LeMessurier and her very red hair as she often has looked up from her kitchen sink and smiled at me. I was pleased with the sight of her. This cul-de-sac led to my family's home, about fifty or so yards of the driveway through our neighbor's property. I took one step onto the driveway.

My family dog, Duke, came running absolutely fucking fast down the driveway. I feared a collision. Hugging and muffling the dog, I approached the house, entering through that basement door that is always unlocked. In the basement, at a trach can, I took off my boots and quietly deposited them in the barrel. Upstairs to the living room, I smelled a strange, tainted odor. There was my brother Steven on a couch, and there was the smell I recognize as Mary Jane.

"Hey, man! How ya doing, Ned?" Steven jumped up and shook my hand, stating, "How was it?"

Sitting down on an opposite couch, I looked in at the room and took a bite of banana and drank a sip of milk that I poured. That made me swoon and said, "I don't know how to tell you this, but," and as I spoke, I thought about monkeys.

ABOUT THE AUTHOR

Ned McFadden graduated from Fordham College. He was an Army medic, drove a taxi, and worked twenty-five years for the federal government. He is the brother of prolific author Steven McFadden and the nephew and godson of Paul Fitzsimmons, the noted author.